T0350191

A Central European Olympiad

The Mathematical Duel

Problem Solving in Mathematics and Beyond

Series Editor: Dr. Alfred S. Posamentier
Chief Liaison for International Academic Affairs
Professor Emeritus of Mathematics Education
CCNY - City University of New York

Long Island University
1 University Plaza -- M101
Brooklyn, New York 11201

There are countless applications that would be considered problem solving in mathematics and beyond. One could even argue that most of mathematics in one way or another involves solving problems. However, this series is intended to be of interest to the general audience with the sole purpose of demonstrating the power and beauty of mathematics through clever problem-solving experiences.

Each of the books will be aimed at the general audience, which implies that the writing level will be such that it will not engulfed in technical language — rather the language will be simple everyday language so that the focus can remain on the content and not be distracted by unnecessarily sophiscated language. Again, the primary purpose of this series is to approach the topic of mathematics problem-solving in a most appealing and attractive way in order to win more of the general public to appreciate his most important subject rather than to fear it. At the same time we expect that professionals in the scientific community will also find these books attractive, as they will provide many entertaining surprises for the unsuspecting reader.

Published

Vol. 7 *A Central European Olympiad: The Mathematical Duel*
by Robert Geretschläger, Józef Kalinowski and Jaroslav Švrček

Vol. 6 *Teaching Children to Love Problem Solving*
by Terri Germain-Williams

Vol. 5 *Strategy Games to Enhance Problem-Solving Ability in Mathematics*
by Alfred S. Posamentier and Stephen Krulik

Vol. 4 *Mathematics Problem-Solving Challenges for Secondary School Students and Beyond*
by David Linker and Alan Sultan

For the complete list of volumes in this series, please visit www.worldscientific.com/series/psmb

Problem Solving in Mathematics and Beyond | Volume 07

A Central European Olympiad

The Mathematical Duel

Robert Geretschläger
BRG Keplerstrasse, Graz, Austria

Józef Kalinowski
University of Silesia in Katowice, Poland

Jaroslav Švrček
Palacký University, Olomouc, Czech Republic

World Scientific

NEW JERSEY • LONDON • SINGAPORE • BEIJING • SHANGHAI • HONG KONG • TAIPEI • CHENNAI • TOKYO

Published by

World Scientific Publishing Co. Pte. Ltd.
5 Toh Tuck Link, Singapore 596224
USA office: 27 Warren Street, Suite 401-402, Hackensack, NJ 07601
UK office: 57 Shelton Street, Covent Garden, London WC2H 9HE

Library of Congress Cataloging-in-Publication Data
Names: Geretschläger, Robert, author. | Kalinowski, Józef, author. | Švrček, Jaroslav, 1953– author.
Title: A Central European Olympiad : the Mathematical Duel / by Robert Geretschläger (BRG Keplerstrasse, Austria), Józef Kalinowski (University of Silesia in Katowice, Poland), Jaroslav Švrček (Palacký University, Olomouc, Czech Republic).
Other titles: Mathematical Duel
Description: New Jersey : World Scientific, 2017. | Series: Problem solving in mathematics and beyond ; volume 7
Identifiers: LCCN 2017037404| ISBN 9789813226166 (hc : alk. paper) | ISBN 9789813223905 (pbk : alk. paper)
Subjects: LCSH: Mathematical recreations. | Mathematics--Competitions--Europe, Central. | Mathematics--Problems, exercises, etc.
Classification: LCC QA95 .G3945 2017 | DDC 510.76--dc23
LC record available at https://lccn.loc.gov/2017037404

British Library Cataloguing-in-Publication Data
A catalogue record for this book is available from the British Library.

For any available supplementary material, please visit
http://www.worldscientific.com/worldscibooks/10.1142/10544#t=suppl

Desk Editors: V. Vishnu Mohan/Tan Rok Ting

Typeset by Stallion Press
Email: enquiries@stallionpress.com

Printed in Singapore

Preface

The world of mathematics competitions has expanded a great deal in the last few decades, and this expansion does not yet seem to have reached an upper limit. It seems that more and more types of competitions are being introduced all over the world to ever more students in ever more diverse circumstances. There are several reasons for this. On the one hand, organizers hope to offer an entertaining and intellectually stimulating diversion to the participants, and at least anecdotally, this certainly seems to succeed in many cases. From a teacher's point of view, there is the hope that a positive emotion connected with the competition experience will be carried over to mathematics in general. On the other hand, higher-level mathematics competitions create an environment that simulates mathematical research, and many academics, both in mathematics and in related scientific fields, have found that participation in math competitions gave them a first insight into what it means to participate actively in mathematical research. Providing such an insight is another important impetus for creating competitions for interested students.

The Mathematical Duel hopes to help participants find their way toward these same positive goals, of course. In addition, the international scope of the school partnerships forming the foundation of the competition is meant to foster international understanding and friendship. In the quarter century since its inception, the political situation in this part of the world has changed a great deal, of course. When the Duel was started, the Iron Curtain, dividing East and West in Europe had just fallen, and regular contacts between schools in Poland, the Czech Republic and Austria still had a bit of an exotic flavor. This is no longer the case, with all three countries in the European Union and travel between the countries flowing freely. Still, for the participating students, participation in the Duel is often their first contact with the cultures of the other countries, and so this aspect is still quite present, even now.

The problems in this set were suggested by the authors of this book as well as several interested colleagues, and the solution texts were produced by the authors, even if the solution ideas may have germinated elsewhere. Perhaps a contest participant had an especially interesting idea, and such an idea may have yielded a specific solution, or perhaps the problem proposer provided the concepts used in the solution presented here. In any case, the authors take full responsibility for any errors that may have found their way into the text. (We do hope that you won't be able to find any, though!). Many mathematical problems allow more than one solution, and multiple solutions are offered for some individual problems. The choice of problems for which solutions are included here was made to reflect the authors' feelings about which will be the most interesting and useful for students preparing for competitions themselves or teachers preparing their students for such events. Any reader is, of course, heartily invited to find solutions to all problems, and there is certainly much left to discover.

Special thanks for their roles in the production of this book go to V. Vishnu Mohan, Alfred Posamentier, and Tan Rok Ting.

Robert Geretschläger
Graz, Austria

About the Authors

Robert Geretschläger is a mathematics teacher at Bundesrealgymnasium Keplerstraße in Graz, Austria. He is also a lecturer at the Karl-Franzens University in Graz, coach and leader of the Austrian International Mathematical Olympiad team and president of the "Kangaroo" student mathematics contest in Austria. He is author of *Geometric Origami*, among other works.

Józef Kalinowski graduated with a PhD in pure mathematics and worked for 27 years as an academic teacher in the Department of Mathematics at the Silesian University of Katowice, Poland. He was also employed at the Institute of Teacher Training in Katowice. Since 1973, he has been a member of the District Committee of the Mathematical Olympiad in Katowice, and has been a vice-chair of this committee since 1981.

Jaroslav Švrček is senior lecturer at Palacký University in Olomouc, Czech Republic. He is Vice-Chair of the Czech Mathematical Olympiad and member of its problem committee. He has been leader or deputy leader of the Czech IMO, MEMO and EGMO teams many times. He is editor of the journal Mathematics Competitions (WFNMC) and also editor for mathematics of the Czech journal Matematika-Fyzika-Informatika.

Contents

Chapter 1
Introduction

A Special Kind of Mathematics Competition

One of the wonderful aspects of international mathematics competitions is the opportunity they offer to bring together people from different geographical backgrounds, but sharing a strong common interest. As we all know, a strong interest in mathematics is not all that common, and young people fascinated by logical thought and abstract puzzles may not always find it easy to find like-minded peers in their own classrooms. As anyone who has ever been involved with any type of international competition can attest to, such a meeting of competitors gives the participants an opportunity to forge friendships across borders that can last a lifetime. Even those taking part who are not quite so fortunate, can at least hope for a chance to gain some understanding of other cultures and other viewpoints. This is the story of an international mathematics competition that has been going strong for nearly a quarter century, and shows no sign of burning out.

Since 1993, the mathematics competition "*Mathematical Duel*" has been held annually between students from two central European schools, the Gymnázium Mikuláše Koperníka (GMK) in Bílovec, Czech Republic, and the I Liceum Ogółnokształcące im. Juliusza Słowackiego (now the Akademicki Zespół Szkół Ogólnokształcących) in Chorzów, Poland. Since 1997, they have been joined by the Bundesrealgymnasium (BRG) Kepler in Graz, Austria, and since 2008 by Gymnázium Jakuba Škody (GJŠ) in Přerov, Czech Republic. The original inspiration for this activity came from several international competitions that were already well established at the time, such as the Baltic Way or the Austrian-Polish Mathematics Competition, but the Duel was to develop into something quite different.

Structure of the Mathematical Duel

Anyone familiar with olympiad-style mathematics will immediately recognise the typical type of problems posed at the Duel. As is the case in many such competitions the world over, participants aim to write mathematical proofs for problems stemming from various topics in pure mathematics and appropriate to their age- and competence-levels.

The competition is divided into three categories. Similar to the Czech olympiad system, division A is for grades 11–12, B for grades 9–10 and C for grades 8 or younger. Typically, four students from each school in each of the divisions come together for a competition (on occasion, a school will not be able to field the complete contingent of four for a group because of sudden illness or for some other reason).

The students participate in an individual competition comprising four problems to be solved in 150 minutes and in a team competition comprising three problems to be solved in 100 minutes. The two competitions are completely independent of one another and yield separate results. While the individual competition is written in supervised silence, the team competition sees the four-member team from each school in each category placed together in a room with no supervision. The students spend their time devising a common group answer to each problem, and only one answer sheet is accepted from each group at the end. The dynamics of the group competition are quite different from the individual work common to most competitions, and are quite fascinating to observe.

The venue for the competition rotates annually between the participating cities Bílovec, Chorzów, Graz and Přerov. In early years, the participants were given the questions in their own languages, but English has been used as a common neutral language since Duel VIII in the year 2000. The students still write their answers in their own languages, and any difficulties in understanding the problems are solved on the spot as the problems are handed out. This method of dealing with the language issue certainly makes for some interesting situations, but the jury has always been able to iron out any difficulties as they have arisen.

The competition was traditionally organised over the course of four days, the first and last of which are spent travelling. (This has changed in the last two years, since the Duel became a part of an EU-sponsored Erasmus Plus project. More will be said about this further on.) The farthest any two of the schools involved are apart is about 800 km, and travel is therefore always possible by bus or train. On one day, there is always an excursion for the participants. This can include some hiking, or a visit to a museum. Also, other activities like sports or puzzle events are organised. Of course, there are also some small prizes for the students with the best results, as well as t-shirts or other souvenirs of the competition, and diplomas for all the participants.

A Special Duel and Guest Schools

In the year 2005, Duel XIII was slated to take place in Graz. Coincidentally, this was the year that BRG Kepler was planning its second Europe Days, a school project meant to foster understanding between the peoples of Europe. It was therefore decided to have a special "European" Duel. Several other schools, with which a long-standing cooperation already existed, were therefore invited to participate in the Mathematical Duel as guests. This was the start of an idea which has become a tradition at the Mathematical Duel, namely for the hosting school to invite partner schools to participate as guests, and this tradition has helped to increase the international flavour of the competition a great deal. The following schools have participated in the Mathematical Duel as guests since 2005:

2005 — Duel XIII — Graz: Diósgyőri Gimnázium (Kilián György Gimnázium), Miskolc, Hungary; GJŠ Přerov;

2006 — Duel XIV — Bílovec: GJŠ Přerov;

2008 — Duel XVII — Přerov: Gymnázium Olomouc — Hejčín, Czech Republic; Slovanské Gymnázium Olomouc, Czech Republic;

2009 — Duel XVII — Graz: mixed team, various schools in Graz, Austria;

2010 — Duel XVIII — Chorzów: Colegiul Naţional "Ion Luca Caragiale", Ploieşti, Romania;

2011 — Duel XIX — Přerov: Liceo Scientifico Statale "Antonio Labriola", Roma-Ostia, Italy;

2012 — Duel XX — Bílovec: Liceo Scientifico Statale "Antonio Labriola", Roma-Ostia, Italy; Colegiul Naţional "Ion Luca Caragiale", Ploieşti, Romania; Sofiyska Matematicheska Gimnaziya Paisiy Hilendarski, Sofia, Bulgaria;

2013 — Duel XXI — Graz: mixed team, various schools in Graz, Austria;

2014 — Duel XXII — Přerov: Sofiyska Matematicheska Gimnaziya Paisiy Hilendarski, Sofia, Bulgaria;

2015 — Duel XXIII — Bielsko-Biała: V Liceum Ogólnokształcące, Bielsko-Biała, Poland.

Reaching Out Beyond the Core — The Mathematical Duel Outside the Participating Schools

Since several schools have participated in the Mathematical Duel as guests along with the regulars, awareness of the competition outside the confines of the partner schools has developed quite naturally. Along with this, however, a conscious effort has also been made to make the problems from the competition available to a wider audience. Articles about the competition were published in various regional publications as well as in the international journal *Mathematics Competitions*, the periodical of the World Federation of National Mathematics Competitions. (*A Local International Mathematics Competition* (Special Edition), Robert Geretschläger and Jaroslav Švrček, Mathematics Competitions, Vol. 18, No. 1, 2005, pp. 39–51.)

Starting in 2008 (with an interruption in 2009), booklets were published containing the complete problems and solutions of the competition, along with the complete results. These booklets were distributed among mathematics educators in the participating countries.

This lead to the use of the problems of the Duel for another, similar competition, annually pitting the students of the BG/BRG Leibnitz (Austria) against those from the Österreichische Schule (Austrian High School) in Budapest (Hungary).

This development has recently found an unexpected climax, with the Mathematical Duel finding Europe-wide recognition.

The Mathematical Duel as an Erasmus Plus Project

Erasmus Plus is a program combining the various funding schemes the European Union finances for education, training, youth and sport. It was launched in its current form as a continuation of numerous predecessor programs in January 2014, and the idea to apply for an Erasmus Plus grant for a project that would include the Duel was born as soon as the creation of the program was announced.

In order to qualify as an Erasmus Plus program, the scope of the activity needed to be widened quite a bit. Most importantly, the Duel needed to be included in a research framework, and this meant widening the group of institutions involved to include some tertiary institutions. For this reason, a university from each of the partner countries was included in the project, now named "*Mathematical Duel plus*". These are The Karl-Franzens University in Graz, Austria, the Palacký University in Olomouc, Czech Republic, and the University of Silesia in Katowice, Poland. A wide research program, looking into the implications of mathematics competitions in education from various viewpoints was introduced, and several academic papers have already appeared. Along with many other documents relating to the project, these can be found at the project website at

http://www.mathematicalduel.eu/.

Since the rules of the Erasmus program require international student activities to be at least five days in length, a fifth day was added to the usual program of the mathematical duel, in which a scientific program with mathematical talks and other activities is organised for the students.

Due to the widened scope of the competition, the competition has been shifted to slightly larger venues in the years of the Erasmus Plus project. In 2015, the contest was held at a hotel catering specially to sports teams in Bialsko-Biała (which is not too far from Chorzów), and in 2016 it was held at a hotel in Ostrava (a larger city near Bílovec).

The inclusion of the Mathematical Duel in this Erasmus Plus project (as this is being written, two of the three years of the initial project have passed) can certainly be considered a success by any measure. A great deal of material is being produced for use in schools all over the European Union, the participating students are being given the opportunity to experience much more mathematics and more cultural crossover than would otherwise be the case, and the website makes the material accessible to anyone interested in immersing themselves in the project.

Problem Development

Finally, before we get to the problems themselves, a few words concerning their genesis seem in order.

Finding appropriate problems for the Mathematical Duel is not an easy task. As is the case for so many similar competitions, every effort is made to develop original material that has not appeared in the published problem sets from all the various competitions around the world. Not only that, the problems should be interesting, age-appropriate and similar in style to the type of problem the students are prepared for. How well we have succeeded in this is up to each reader of this collection to decide for him- or herself.

The majority of the problems used in the competition over the years were suggested by the three authors of this book, but many were suggested by our colleagues. Unfortunately, in the early years of the competition, there was not really any thought given to collecting information like the names of the proposers of specific problems for posterity. For this reason, we must apologise for possibly not being able to thank some early collaborator for his or her contributions. Since 2011, however, this information has been collected systematically, and together with the names of problem proposers we were able to reconstruct from various other sources, we gratefully acknowledge the following problem authors for their suggestions:

Pavel Calábek (A-T-1-99, A-T-1-04, A-T-3-07, A-T-3-08, A-I-1-09, A-T-3-11, C-I-4-11, A-I-3-12, A-T-1-12, A-T-3-12, A-I-2-13, A-T-2-13, A-I-2-14, A-T-2-14, A-T-1-15, B-T-3-15, C-T-1-15, A-I-4-16, A-T-3-16, B-I-4-16)

Emil Calda (A-T-I-02)

Radek Horenský (A-I-4-10)

Gottfried Perz (A-I-4-15, B-I-2-15)

Stanislav Trávníček (B-I-1-14, C-I-4-14)

Jacek Uryga (A-I-3-11, A-T-1-11, B-T-2-11, A-I-1-12, B-I-2-12, A-I-1-13, B-I-3-13, C-I-2-13, A-I-1-14, A-I-3-14, B-I-3-14, A-I-2-15, A-T-2-15, C-T-2-15, A-T-1-16, A-T-2-16, B-I-3-16, B-T-3-16)

Erich Windischbacher (C-T-2-12, A-I-3-13)

Note that each problem is referred to in the form (category–individual/team–number–year). This means that (B-I-2-03) refers to problem number 2 of the individual competition in the B category of the year

2003. Furthermore, the number of the competition results by adding 8 to the digits of the year. The year 2003 was therefore the year of the $03+8 = 11$th Duel.

In each subsection, a brief introduction is followed by a list of problems. Interested readers are invited to try their hand at finding solutions before heading on to the solutions that follow immediately afterwards. Of course, if you prefer to head straight to the solutions, you may just want to skip right ahead. Either way, we hope you will enjoy this collection of problems from our Mathematical Duel.

Chapter 2

Number Theory

2.1. Numbers with Interesting Digits

A type of problem used quite often in the Mathematical Duel concerns problems that specifically address numbers with unusual digits. This can mean numbers whose digits are all the same, or come from some limited set, or numbers in which the digits have some unusual property, as is the case with palindromic numbers. None of these problems is particularly difficult, but the level of difficulty in the collection is surprisingly varied. Tools required to solve the problems come from a surprising variety of sources, considering how similar the questions appear to be on a superficial first reading. Beside number theory, we use some tools from combinatorics and algebra, such as the pigeon-hole principle or induction.

PROBLEMS

1. We consider positive integers that are written in decimal notation using only one digit (possibly more than once), and call such numbers *uni-digit numbers*.

(a) Determine a uni-digit number written with only the digit 7 that is divisible by 3.
(b) Determine a uni-digit number written with only the digit 3 that is divisible by 7.
(c) Determine a uni-digit number written with only the digit 5 that is divisible by 7.
(d) Prove that there cannot exist a uni-digit number written with only the digit 7 that is divisible by 5.

2. We call a number that is written using only the digit 1 in decimal notation a *onesy* number, and a number using only the digit 7 in decimal notation a *sevensy* number. Determine a onesy number divisible by 7 and prove that for any sevensy number k, there always exists a onesy number m such that m is a multiple of k.

3. (a) A number x can be written using only the digit a both in base 8 and in base 16, i.e.,

$$x = (aa\dots a)_8 = (aa\dots a)_{16}.$$

Determine all possible values of x.

(b) Determine as many numbers x as possible that can be written in the form $x = (11\dots 1)_b$ in at least two different number systems with bases b_1 and b_2.

4. We call positive integers that are written in decimal notation using only the digits 1 and 2 *Graz numbers*. Note that 2 is a 1-digit Graz number divisible by 2^1, 12 is a 2-digit Graz number divisible by 2^2 and 112 is a 3-digit Graz number divisible by 2^3.

(a) Determine the smallest 4-digit Graz number divisible by 2^4.
(b) Determine an n-digit Graz number divisible by 2^n for $n > 4$.
(c) Prove that there must always exist an n-digit Graz number divisible by 2^n for any positive integer n.

5. Determine the number of ten-digit numbers divisible by 4 which are written using only the digits 1 and 2.

6. Let A be a six-digit positive integer which is formed using only the two digits x and y. Furthermore, let B be the six-digit integer resulting from A if all digits x are replaced by y and simultaneously all digits y are replaced by x. Prove that the sum $A + B$ is divisible by 91.

7. Determine the number of pairs (x, y) of decimal digits such that the positive integer in the form $\overline{xyx}$ is divisible by 3 and the positive integer in the form $\overline{yxy}$ is divisible by 4.

8. Determine the number of all six-digit palindromes that are divisible by seven.

Remark. A six-digit palindrome is a positive integer written in the form $\overline{abccba}$ with decimal digits $a \neq 0$, b and c.

9. Two positive integers are called *friends* if each is composed of the same number of digits, the digits in one are in increasing order and the digits in the other are in decreasing order, and the two numbers have no digits in common (like for example the numbers 147 and 952).

Solve the following problems:

(a) Determine the number of all two-digit numbers that have a friend.
(b) Determine the largest number that has a friend.

10. A *wavy number* is a number in which the digits alternately get larger and smaller (or smaller and larger) when read from left to right. (For instance, 3629263 and 84759 are wavy numbers but 45632 is not.)

(a) Two five-digit wavy numbers m and n are composed of all digits from 0 to 9. (Note that the first digit of a number cannot be 0.) Determine the smallest possible value of $m + n$.
(b) Determine the largest possible wavy number in which no digit occurs twice.
(c) Determine a five-digit wavy number that can be expressed in the form $ab + c$, where a, b and c are all three-digit wavy numbers.

SOLUTIONS

Problem 1. (C-T-2-13) *We consider positive integers that are written in decimal notation using only one digit (possibly more than once), and call such numbers* uni-digit numbers.

(a) *Determine a uni-digit number written with only the digit* 7 *that is divisible by* 3.
(b) *Determine a uni-digit number written with only the digit* 3 *that is divisible by* 7.
(c) *Determine a uni-digit number written with only the digit* 5 *that is divisible by* 7.
(d) *Prove that there cannot exist a uni-digit number written with only the digit* 7 *that is divisible by* 5.

Solution: Part (a) is easily solved either by trying out small numbers of this type, or by noting that a number is divisible by 3 if the sum of its digits is. We therefore have $777 = 7 \cdot 111 = 7 \cdot 37 \cdot 3$.

(b) Here, we can once again try to simply divide candidate numbers like 3, 33, 333 and so on by 7. On the other hand, we can note that $1001 = 7 \cdot 11 \cdot 13$, and therefore $333333 = 333 \cdot 1001 = 333 \cdot 7 \cdot 11 \cdot 13$.
(c) As for (b), we get $555555 = 555 \cdot 7 \cdot 11 \cdot 13$.
(d) The last digit of any number divisible by 5 is always either 0 or 5. Any number that is divisible by 5 can therefore not be written using only the digit 7. □

Problem 2. (B-I-4-13) *We call a number that is written using only the digit* 1 *in decimal notation a* onesy *number, and a number using only the digit* 7 *in decimal notation a* sevensy *number. Determine a onesy number divisible by* 7 *and prove that for any sevensy number* k*, there always exists a onesy number* m *such that* m *is a multiple of* k*.*

Solution: Onesy numbers are of the form $111\ldots111$. We easily see that the smallest of them are not divisible by 7: 1; 11; $111 = 3 \cdot 37$; $1111 = 11 \cdot 101$. By continuing this experimentation, or by the same reasoning as in the previous problem, we see that a possible onesy number divisible by seven is given by $111111 = 111 \cdot 1001 = 111 \cdot 7 \cdot 11 \cdot 13$.

Sevensy numbers are of the form $k = 777\ldots777$. In order to see that there always exists a onesy multiple of any sevensy number k, we note that there exist an infinite number of onesy numbers. By the Dirichlet (pigeon-hole) principle, there must therefore exist two different onesy numbers $m_1 > m_2$ with $m_1 \equiv m_2 \pmod{k}$. Writing $m_1 = 11111\ldots111$ and $m_2 = 111\ldots111$, we therefore have $m_1 - m_2 = 11\cdots1100\ldots00$.

It therefore follows that $m_1 - m_2$ is divisible by k. The number $m_1 - m_2$ can be written as $m_1 - m_2 = m \cdot 10^r$, where m is also a onesy number. Since k is certainly not divisible by 2 or 5, it follows that m must also be divisible by k, and the proof is complete. □

Problem 3. (B-T-1-05)

(a) *A number* x *can be written using only the digit* a *both in base* 8 *and in base* 16*, i.e.,*

$$x = (aa\ldots a)_8 = (aa\ldots a)_{16}.$$

Determine all possible values of x*.*

(b) *Determine as many numbers x as possible that can be written in the form $x = (11\ldots1)_b$ in at least two different number systems with bases b_1 and b_2.*

Solution: (a) If $(aa\ldots a)_8 = (aa\ldots a)_{16}$ holds, there exist m and n such that

$$\begin{aligned} &a \cdot 16^m + a \cdot 16^{m-1} + \cdots + a \cdot 16 + a \\ &= a \cdot 8^n + a \cdot 8^{n-1} + \cdots + a \cdot 8 + a \end{aligned}$$

holds.
This is equivalent to

$$\begin{aligned} &16^m + \cdots + 16 = 8^n + \cdots + 8 \\ \iff &16 \cdot \frac{16^m - 1}{16 - 1} = 8 \cdot \frac{8^n - 1}{8 - 1} \\ \iff &2 \cdot \frac{16^m - 1}{15} = \frac{8^n - 1}{7} \\ \iff &\frac{2 \cdot 16^m - 2}{15} = \frac{8^n - 1}{7} \\ \iff &14 \cdot 16^m - 14 = 15 \cdot 8^n - 15 \\ \iff &15 \cdot 8^n = 14 \cdot 16^m + 1. \end{aligned}$$

The right side is odd. Therefore, we have $n = 0$ and $m = 0$.
The only possible values of a are $a \in \{0, 1, 2, \ldots, 7\}$, and we therefore have $x \in \{0, 1, 2, \ldots, 7\}$.

(b) If $x = (11\ldots1)_{b_1} = (11\ldots1)_{b_2}$, we have $x = 1$ or $b_1, b_2 > 1$.
Assume $1 < b_1 < b_2$. We want

$$x = \sum_{i=0}^{m} b_1^i = \sum_{j=0}^{n} b_2^j \quad \text{with } m > n.$$

For any $b_1 > 1$ choose $b_2 = \sum_{i=1}^{m} b_1^i$. Then

$$(11)_{b_2} = 1 \cdot b_2 + 1 \cdot b_2^0 = \sum_{i=1}^{m} b_1^i + 1 \cdot b_1^0 = \sum_{i=0}^{m} 1 \cdot b_1^i = (11\ldots1)_{b_1},$$

and we have infinitely many x with the required property. □

Problem 4. (A-T-3-13) *We call positive integers that are written in decimal notation using only the digits* 1 *and* 2 Graz numbers. *Note that* 2 *is a* 1*-digit Graz number divisible by* 2^1, 12 *is a* 2*-digit Graz number divisible by* 2^2 *and* 112 *is a* 3*-digit Graz number divisible by* 2^3.

(a) *Determine the smallest* 4*-digit Graz number divisible by* 2^4.
(b) *Determine an* n*-digit Graz number divisible by* 2^n *for* $n > 4$.
(c) *Prove that there must always exist an* n*-digit Graz number divisible by* 2^n *for any positive integer* n.

Solution: (a) Some 4-digit candidates are 2222, 1212, 2112. Prime decomposition of these numbers gives us

$$2222 = 2 \cdot 1111, \quad 1212 = 2^2 \cdot 303, \quad 2112 = 64 \cdot 33 = 2^6 \cdot 33,$$

and 2112 is therefore a 4-digit Graz number.

(b) and (c) We can prove by induction that there in fact exists an n-digit Graz number for any positive integer n. Obviously, 2 is the only 1-digit Graz number, as 1 is not divisible by 2^1, but 2 is. We can therefore assume that there exists a k-digit Graz number g for some $k \geq 1$. Since g is divisible by 2^k, either $g \equiv 0 \pmod{2^{k+1}}$ or $g \equiv 2^k \pmod{2^{k+1}}$ must hold. Since $10^k \equiv 2^k \pmod{2^{k+1}}$ and $2 \cdot 10^k \equiv 0 \pmod{2^{k+1}}$, we have either $10^k + g \equiv 0 \pmod{2^{k+1}}$ or $2 \cdot 10^k + g \equiv 0 \pmod{2^{k+1}}$, and therefore the existence of an $(n-1)$-digit Graz number.

It is now easy to complete the solution. Since 112 is a 3-digit Graz number, and $112 = 16 \cdot 7$ is divisible by 16, 2112 is a 4-digit Graz number. Since $2112 = 32 \cdot 66$ is divisible by $2^5 = 32$, 22112 is a 5-digit Graz number, and the solution is complete. □

Problem 5. (C-I-3-11) *Determine the number of ten-digit numbers divisible by* 4 *which are written using only the digits* 1 *and* 2.

Solution: An example of such a number is $2212211112 = 4 \cdot 553052778$.

Any ten-digit number n divisible by 4 must end in a two-digit number divisible by 4. The last two digits of any such number written only with the digits 1 and 2 can therefore only be 12, in this order. Each of the eight other eight digits of the ten-digit number can be either 1 or 2. Altogether, this gives us 2^8 possibilities. There therefore exist $2^8 = 256$ ten-digit numbers with the given property. □

Problem 6. (B-I-1-11) *Let A be a six-digit positive integer which is formed using only the two digits x and y. Furthermore, let B be the six-digit integer resulting from A if all digits x are replaced by y and simultaneously all digits y are replaced by x. Prove that the sum $A + B$ is divisible by* 91.

Solution: An example of such a pair A and B is $229299 + 992922 = 1222221 = 91 \cdot 13431$. The claim certainly holds in this particular case.

In order to prove the general assertion, let $A = \overline{c_5c_4c_3c_2c_1c_0}$ and $B = \overline{d_5d_4d_3d_2d_1d_0}$, where $c_i, d_i \in \{x, y\}$, $c_i \neq d_i$ for $i = 0, 1, 2, 3, 4, 5$ and $x, y \in \{1, \dots, 9\}$ are distinct non-zero decimal digits.
Since $c_i + d_i = x + y \neq 0$ for $i = 0, 1, 2, 3, 4, 5$ we have

$$\begin{aligned} A + B &= (x + y) \cdot (10^5 + 10^4 + 10^3 + 10^2 + 10 + 1) \\ &= (x + y) \cdot 111111 = (x + y) \cdot 91 \cdot 1221, \end{aligned}$$

The number $A + B$ is therefore certainly divisible by 91. □

Problem 7. (C-T-1-10) *Determine the number of pairs (x, y) of decimal digits such that the positive integer in the form $\overline{xyx}$ is divisible by* 3 *and the positive integer in the form $\overline{yxy}$ is divisible by* 4.

Solution: An example of such a pair of numbers is given by $525 = 3 \cdot 175$ and $252 = 4 \cdot 63$.

Each positive integer in the form $\overline{yxy}$ is divisible by 4 if and only if the number $\overline{xy}$ is divisible by 4 with $y \neq 0$. Hence

$$\begin{aligned} (x, y) \in \{&(1; 2), (1; 6), (2; 4), (2; 8), (3; 2), (3; 6), (4; 4), (4; 8), \\ &(5; 2), \dots, (5; 6), (6; 4), (6; 8), (7; 2), (7; 6), (8; 4), \\ &(8; 8), (9; 2), (9; 6)\}. \end{aligned}$$

A positive integer in the form $\overline{xyx}$ is divisible by 3 if and only if the sum of its digits is divisible by 3, i.e., $2x + y$ must be divisible by 3. After checking all possible pairs of positive integers we obtain only six possibilities:

$$(x, y) \in \{(2; 8), (3; 6), (4; 4), (5; 2), (8; 8), (9; 6)\}.$$

We therefore have six solutions altogether. □

Problem 8. (B-T-3-15) *Determine the number of all six-digit palindromes that are divisible by seven.*

Remark. A six-digit palindrome is a positive integer written in the form $\overline{abccba}$ *with decimal digits* $a \neq 0$, b *and* c.

Solution: We have

$$\overline{abccba} = 100001a + 10010b + 1100c$$

$$= 7(14286a + 1430b + 157c) - (a - c).$$

Such a number is divisible by 7 if and only if $(a - c)$ is divisible by 7. $a \neq 0$ and c are decimal digits, and we therefore have $-8 \leq a - c \leq 9$. The only possible values are therefore $(a - c) \in \{-7, 0, 7\}$.

For $a - c = -7$ we have $(a, c) \in \{(1, 8), (2, 9)\}$, for $a - c = 0$ we have $(a, c) \in \{(1, 1), (2, 2), \ldots, (9, 9)\}$, and finally for $a - c = 7$ we have $(a, c) \in \{(7, 0), (8, 1), (9, 2)\}$, and therefore we have 14 possibilities for the ordered pair (a, c) in total.

In all cases b is an arbitrary digit, and altogether there are therefore $14 \cdot 10 = 140$ six-digit palindromes which are divisible by 7. □

Problem 9. (C-I-3-12) *Two positive integers are called friends if each is composed of the same number of digits, the digits in one are in increasing order and the digits in the other are in decreasing order, and the two numbers have no digits in common (like for example the numbers* 147 *and* 952).

Solve the following problems:

(a) *Determine the number of all two-digit numbers that have a friend.*
(b) *Determine the largest number that has a friend.*

Solution: (a) Every two-digit number n which is composed of different digits has its digits in increasing or decreasing order. Moreover there are at least two non-zero digits a and b different from the digits of n. It follows that the friend of n is one of numbers $\overline{ab}$ or $\overline{ba}$.

The number of two-digit numbers with a friend is therefore equal to the number of two-digit numbers composed of different digits. There are 90 two-digit numbers of which 9 $(11, 22, \dots, 99)$ consist of two identical digits. There are therefore 81 two-digit numbers which have a friend.

(b) If the number with a friend has k digits, its friend also has k different digits and together they have $2k$ different digits. Since there are 10 digits, the largest number with a friend has at most five digits.

No number begins with 0, so 0 is in a number with digits in decreasing order if $k = 5$. Moreover, if a number n with digits in increasing order has a friend k, its mirror image (that is, the number with the same digits in opposite order) is greater and has a friend (namely the mirror image of k).

The largest number with a friend has different digits in decreasing order, it has at most five digits and one of its digits is 0. The largest such number is therefore 98760 and its friend is 12345. □

Problem 10. (C-T-3-15) *A wavy number is a number in which the digits alternately get larger and smaller (or smaller and larger) when read from left to right. (For instance,* 3629263 *and* 84759 *are wavy numbers but* 45632 *is not.)*

(a) *Two five-digit wavy numbers m and n are composed of all digits from* 0 *to* 9. *(Note that the first digit of a number cannot be* 0.*) Determine the smallest possible value of $m + n$.*
(b) *Determine the largest possible wavy number in which no digit occurs twice.*
(c) *Determine a five-digit wavy number that can be expressed in the form $ab + c$, where a, b and c are all three-digit wavy numbers.*

Solution: (a) The smallest possible sum is given by the expression

$$20659 + 14387 = 35046.$$

(b) The largest such number is 9785634120.

(c) There are many such combinations. Examples are

$$120 \cdot 142 + 231 = 17271 \quad \text{or} \quad 101 \cdot 101 + 101 = 10302. \qquad \square$$

2.2. Integer Fractions

Quite a common theme, especially for easier competition problems, concerns structures that appear to be fractions, but reduce to integers under certain circumstances. Not all of these questions are so easy, though. All manner of basic number theoretical reasoning is needed to be able to deal with these problems, and it is not always obvious at first reading, which of these problems we will be able to breeze through, and which will require some deeper insight.

PROBLEMS

11. Determine all integers n such that the fraction

$$\frac{8n-1}{11n-2}$$

is reducible.

12. Show that

$$n = \frac{2008^3 + 2007^3 + 3 \cdot 2008 \cdot 2007 - 1}{2009^2 + 2008^2 + 1}$$

is an integer and determine its value.

13. Show that

$$m = \frac{2008^4 + 2008^2 + 1}{2008^2 + 2008 + 1}$$

is an integer and determine its value.

14. Determine all positive integers n ($n \leq 200$) such that there exist exactly six integers x for which

$$\frac{n+x}{2-x}$$

is an integer.

15. Determine all positive integers n ($n \leq 50$) such that there exist eight integers x for which

$$\frac{x^2 + 7x - n}{x + 2}$$

is an integer.

16. Determine all integers x such that

$$f(x) = \frac{x^3 - 2x^2 - x + 6}{x^2 - 3}$$

is an integer.

17. Prove that the fraction

$$\frac{n^4 + 4n^2 + 3}{n^4 + 6n^2 + 8}$$

is not reducible for any integer n.

18. Determine all positive integers n such that

$$\frac{19n + 17}{7n + 11}$$

is an integer.

SOLUTIONS

Problem 11. (B-I-1-14) *Determine all integers n such that the fraction*

$$\frac{8n - 1}{11n - 2}$$

is reducible.

Solution: It is easy to see that the denominator of the given fraction is non-zero for all integer values of n.

Let d be the greatest common divisor of the two integers $8n - 1$ and $11n - 2$. Then d also divides the number $11(8n - 1) - 8(11n - 2) = 5$. This implies $d \in \{1; 5\}$. The given fraction is therefore reducible if and only if the number 5 divides both integers $8n - 1$ and $11n - 2$.

Since $8n-1 = 8(n-2)+15$, we have $n-2 = 5k$ for arbitrary integers k, i.e., $n = 5k + 2$. We can easily check that for such n the denominator $11n - 2 = 55k + 20$ of the given fraction is divisible by 5, as well.

We see that the given fraction is reducible (by the number 5) for any $n = 5k + 2$ with arbitrary integer values of k. □

Problem 12. (A-I-1-08) *Show that*

$$n = \frac{2008^3 + 2007^3 + 3 \cdot 2008 \cdot 2007 - 1}{2009^2 + 2008^2 + 1}$$

is an integer and determine its value.

Solution: Substituting $x = 2008$, we obtain

$$n = \frac{x^3 + (x-1)^3 + 3x(x-1) - 1}{(x+1)^2 + x^2 + 1} = \frac{2x^3 - 2}{2x^2 + 2x + 2} = x - 1,$$

and therefore $n = 2008 - 1 = 2007$.

Problem 13. (B-I-1-08) *Show that*

$$m = \frac{2008^4 + 2008^2 + 1}{2008^2 + 2008 + 1}$$

is an integer and determine its value.

Solution: Noting that

$$(x^4 + x^2 + 1) : (x^2 + x + 1) = x^2 - x + 1,$$

we only need to substitute $x = 2008$ to obtain $m = 4030057$.

Problem 14. (B-T-3-03) *Determine all positive integers n ($n \leq 200$) such that there exist exactly six integers x for which*

$$\frac{n + x}{2 - x}$$

is an integer.

Solution: We first note that

$$\frac{n+x}{2-x} = -1 + \frac{n+2}{2-x}.$$

We can therefore consider divisors of the number $n+2$.

If the number $n+2$ is the product of two or more different prime numbers, it certainly has more than six divisors.

If the number $n+10$ is equal to p, where p is a prime number, it has only four divisors: $\pm 1, \pm p$.

If the number $n+10$ is of the form p^2, where p is a prime number, it has six divisors: $\pm 1, \pm p, \pm p^2$.

If the number $n+10$ is of the form p^u, where p is a prime number and u is an integer with $u \geq 3$, it has more than six divisors: $\pm 1, \pm p, \pm p^2, \pm p^3, \ldots$.

Because $n \leq 200$, we require $n+2 \leq 202$ and therefore $p^2 \leq 202$. This gives us the possible values $p \in \{2, 3, 5, 7, 11, 13\}$. Since $n = p^2 - 2$, we therefore obtain exactly six solutions for $n \leq 200$, namely $n \in \{2, 7, 23, 47, 119, 187\}$. □

Problem 15. (A-T-3-03) *Determine all positive integers n ($n \leq 50$) such that there exist eight integers x for which*

$$\frac{x^2 + 7x - n}{x+2}$$

is an integer.

Solution: This is, of course, a slightly more difficult version of the previous problem, and the solution steps will be quite similar. We first note that

$$\frac{x^2 + 7x - n}{x+2} = x + 5 - \frac{n+10}{x+2}.$$

We can therefore consider divisors of the number $n+10$.

If the number $n+10$ is the product of powers of three or more different prime numbers, it certainly has more than eight divisors.

If the number $n+10$ is the product $p \cdot q$ of two different prime numbers p, q, it has exactly eight divisors: $\pm 1, \pm p, \pm q, \pm pq$.

If the number $n+10$ is the product of the powers of two different prime numbers p, q, in the form $p^u \cdot q^v$, where u, v are positive integers with $u + v \geq 3$, the number has more than eight divisors: $\pm 1, \pm p, \pm q, \pm pq, \pm p^2 q$ or $\pm pq^2, \ldots$.

If the number $n + 10$ is of the form p^3, where p is a prime number, it again has eight divisors: $\pm 1, \pm p, \pm p^2, \pm p^3$.

If the number $n + 10$ is of the form p^u, where p is a prime number and u is an integer with $u > 3$, it has more than eight divisors.

If the number $n + 10$ is of the form p^2, where p is a prime number, it has only six divisors.

Using these observations, let us consider two possible cases:

Case 1: $n + 10 = p^3$. Then $n = p^3 - 10$ and since $n \leq 50$, only $p = 3$, is possible, which gives us $n = 17$.

Case 2: $n + 10 = p \cdot q$, where $p \neq q$ are different prime numbers. Since $n \leq 50$, we have $p \cdot q - 10 \leq 50$, and therefore $p \cdot q \leq 60$. Let $p < q$. Because p is not less than 2, q is not greater than 30. The following prime numbers lie between 2 and 30: 2, 3, 5, 7, 11, 13, 17, 19, 23 and 29. Using these numbers, we have 17 products $p \cdot q$ not greater than 60: 6, 10, 14, 15, 21, 22, 26, 33, 34, 35, 38, 39, 46, 51, 55, 57, 58. This yields the following possible values for $n = p \cdot q - 10$ (after we eliminate the non-positive numbers -4 and 0): $n \in \{4, 5, 11, 12, 16, 23, 24, 25, 28, 29, 36, 41, 45, 47, 48\}$. We see that there are 15 solutions for $n \leq 50$. □

Problem 16. (B-I-1-01) *Determine all integers x such that*

$$f(x) = \frac{x^3 - 2x^2 - x + 6}{x^2 - 3}$$

is an integer.

Solution: Because $x^2 \neq 3$ holds for any integer,

$$\frac{x^3 - 2x^2 - x + 6}{x^2 - 3} = x - 2 + \frac{2x}{x^2 - 3}$$

is an integer if and only if $(x^2 - 3) | 2x$. This is only possible if $|x^2 - 3| \leq |2x|$. The only integers x with this property are $x \in \{-3, -2, -1, 0, 1, 2, 3\}$.

For these numbers, we have

$$f(-3) = \frac{-36}{6} = -6, \quad f(-2) = \frac{-8}{1} = -8$$

$$f(-1) = \frac{4}{-2} = -2, \quad f(0) = \frac{6}{-3} = -2, \quad f(1) = \frac{4}{-2} = -2,$$

$$f(2) = \frac{4}{1} = 4 \quad \text{and} \quad f(3) = \frac{12}{6} = 2.$$

We see that the integers x for which $f(x)$ is an integer are the elements of the set $\{-3, -2, -1, 0, 1, 2, 3\}$. □

Problem 17. (B-I-1-97) *Prove that the fraction*

$$\frac{n^4 + 4n^2 + 3}{n^4 + 6n^2 + 8}$$

is not reducible for any integer n.

Solution: We first note that

$$\frac{n^4 + 4n^2 + 3}{n^4 + 6n^2 + 8} = \frac{(n^2 + 1)(n^2 + 3)}{(n^2 + 2)(n^2 + 4)}.$$

The four factors in the numerator and denominator of this fraction are consecutive positive integers. Because two consecutive positive integers k and $k + 1$ are relatively prime, the fraction can only be reducible if $n^2 + 1$ and $n^2 + 4$ have a common factor greater than 1. If $d|n^2 + 1$ and $d|n^2 + 4$, then $d|((n^2 + 4) - (n^2 + 1)) = 3$. Therefore, if the fraction is reducible, then the common divisor d of $n^2 + 1$ and $n^2 + 4$ must be equal to 3. We will now prove that neither of these numbers can be divisible by 3.

If $n \equiv \pm 1 \pmod 3$, then

$$n^2 \equiv 1 \quad \text{and} \quad n^2 + 1 \equiv n^2 + 4 \equiv 2 \pmod 3;$$

and if $n \equiv 0 \pmod 3$, then

$$n^2 \equiv 0 \quad \text{and} \quad n^2 + 1 \equiv n^2 + 4 \equiv 1 \pmod 3.$$

In both cases, we see that neither $n^2 + 1$ nor $n^2 + 4$ can be divisible by 3, and this completes our proof. □

Problem 18. (B-T-3-94) *Determine all positive integers n such that*

$$\frac{19n + 17}{7n + 11}$$

is an integer.

Solution: We first note that

$$\frac{19n + 17}{7n + 11} = 2 + \frac{5n - 5}{7n + 11}$$

holds for any positive integer n.

For $n = 1$, the fraction has the integer value equal to 2.

For $n > 1$, the inequality $5n - 5 < 7n + 11$ holds, and the fraction

$$0 < \frac{5n - 5}{7n + 11} < 1$$

cannot assume an integer value. We see that the fraction cannot assume integer values for any $n > 1$. □

Another solution: Let us define a sequence by

$$a_n = \frac{19n + 17}{7n + 11}$$

for positive integers n. The terms of this sequence $\{a_n\}$ are certainly all positive. We shall prove that the sequence is increasing. For this purpose, we consider the difference

$$a_{n+1} - a_n = \frac{19(n + 1) + 17}{7(n + 1) + 11} - \frac{19n + 17}{7n + 11} = \frac{90}{(7n + 11)(7n + 18)} > 0$$

of successive terms of the sequence. Since this difference is always positive, the sequence is certainly increasing.

We note that $a_1 = 2$. Next, we can show that $a_n < 3$ holds for all positive integers n. Let us assume that $a_n \geq 3$ holds for some n. Then we have

$$\frac{19n + 17}{7n + 11} \geq 3,$$

which is equivalent to $n \leq -8$, yielding a contradiction. We see that the inequalities $2 \leq a_n < 3$ hold for all positive integers n. Since a_n is an integer, the only possibility is $a_n = 2$, and since the sequence $\{a_n\}$ is increasing, this is only possible for $n = 1$. □

Another solution: As in the previous solution, we define the a_n for positive integers n and show that $a_n < 3$ must hold.

The only possible integer values of the fraction a_n are therefore 1 or 2. Since the equation $\frac{19n+17}{7n+11} = 1$ has the solution $n = -\frac{1}{2}$ and the equation $\frac{19n+17}{7n+11} = 2$ has the solution $n = 1$, we see that the value of the fraction can only be an integer for $n = 1$. □

2.3. Divisibility

The bread and butter of number theory is the question of divisibility. It is only natural that many competition problems in number theory ask questions of this elementary type.

PROBLEMS

19. Find the largest positive integer n with the following property: The product

$$(k+1) \cdot (k+2) \cdot (k+3) \cdot \ldots \cdot (k+2016)$$

is divisible by 2016^n for every positive integer k.

20. Prove that for every positive integer n, there exists an integer m divisible by 5^n which consists exclusively of odd digits.

21. Consider the numbers from the set $\{1, 2, 3, \ldots, 2016\}$. How many of these have the property that its square leaves a remainder of 1 after division by 2016?

22. Determine the number of all six-digit palindromes which are divisible by seven.

23. We are given seven distinct positive integers. Prove that four of these can be chosen such that their sum is divisible by four.

24. Determine all triples (x, y, z) of positive integers for which each of the three numbers x, y, z is a divisor of the sum $x + y + z$.

25. Peter throws two dice simultaneously and then writes the number of dots showing on a blackboard. Find the smallest number k with the following property: After k throws Peter can always choose some of the written numbers such that their product leaves the remainder 1 after division by 13.

26. How many positive integers of the form $\overline{abcabcabc}$ exist that are divisible by 29?

27. Prove that the number $2010^{2011} - 2010$ is divisible by $2010^2 + 2011$.

28. Determine the largest positive integer k such that the number $n^6 - n^4 - n^2 + 1$ is divisible by 2^k for every odd integer $n > 1$.

29. Determine all three-digit numbers $\overline{xyz}$ which are divisible by all three digits x, y, z and with no two digits being equal.

30. Prove that the expression $2^{12n+8} - 3^{6n+2}$ is divisible by 13 for every non-negative integer value n.

31. Prove that

$$L = m^{21}n^3 - m^7n - m^3n^{21} + mn^7$$

is divisible by 42 for all integer values of m and n.

32. Prove that the number $m^5n - mn^5$ is divisible by 30 for all positive integers m and n.

33. Prove that a two-digit number, which is not divisible by 10 but is divisible by the sum of its digits, is certainly divisible by 3. Is this also true for all three-digit numbers with this property?

34. We are given the set of all seven-digit numbers, in which the digits are some permutation of $1, 2, 3, \ldots, 7$. Prove that the sum of all these numbers is divisible by 9.

35. Prove that the number

$$1993^2 - 1992^2 + 1991^2 - 1990^2 + \ldots + 3^2 - 2^2 + 1^2$$

is divisible by 1993.

SOLUTIONS

Problem 19. (A-I-1-16) *Find the largest positive integer n with the following property: The product*

$$(k+1) \cdot (k+2) \cdot (k+3) \cdot \ldots \cdot (k+2016)$$

is divisible by 2016^n *for every positive integer k.*

Solution: The prime decomposition of 2016 is given by $2016 = 2^5 \cdot 3^2 \cdot 7$. It is easy to see that every other factor in the product is divisible by 2, and similarly every third factor by 3 and every seventh factor by 7. Since every fourth factor is even divisible by 2^2, we need not deal with powers of 2 or 3. If we want to find the largest possible positive integer n with the given property, we only need to find the largest n such that 7^n is a divisor of the given product for every positive integer k. Thus, the problem reduces to finding that n.

Since every seventh factor of the given product is divisible by 7, we have $2016{:}7 = 288$ numbers divisible at least by 7. Moreover, every 49th factor is divisible by $49 = 7^2$. Thus, we altogether have $[2016{:}48] = 41$ (we use $[x]$ to denote the largest integer not greater than a real number x). Finally, every 343rd factor is divisible by $343 = 7^3$, and we certainly have another $[2016{:}343] = 5$ such factors of 7 contained in the product. (Note that we do not need to consider higher powers of 7, since $7^4 > 2016$ holds.)

It therefore follows that $n = 288 + 41 + 5 = 334$ sevens are certainly contained in the product, completing the proof. □

Problem 20. (A-T-1-16) *Prove that for every positive integer n, there exists an integer m divisible by 5^n which consists exclusively of odd digits.*

Solution: We prove a stronger assumption: For every positive integer n, there exists an integer m divisible by 5^n which consists exclusively of n odd digits.

Note first that for every positive integer n and q the numbers $2^n + q$, $3 \cdot 2^n + q, 5 \cdot 2^n + q, 7 \cdot 2^n + q$ and $9 \cdot 2^n + q$ give five different remainders when divided by 5. Indeed, if two of these remainders were equal, then the difference of those numbers would be divisible by 5, which is certainly not possible.

For $n = 1$ we can choose $m = 5$. Now suppose, for some n we have a number m divisible by 5^n consisting of n odd digits.
Let us take $q = \frac{m}{5^n}$ and choose $d \in \{1, 3, 5, 7, 9\}$ such that $5|d \cdot 2^n + q$. In this case, the number

$$m' = 5^n(d \cdot 2^n + q) = d \cdot 10^n + m$$

is certainly divisible by 5^{n+1}. The number m' is therefore composed of $n+1$ odd digits, namely the n odd digits of m and the odd digit d. The existence of m' therefore completes our proof. □

Problem 21. (B-I-4-16) *Consider the numbers from the set* $\{1, 2, 3, \ldots, 2016\}$. *How many of these have the property that its square leaves a remainder of* 1 *after division by* 2016?

Solution: We first note that the prime decomposition of 2016 is given by $2016 = 2^5 \cdot 3^2 \cdot 7$. The remainder of an integer x^2 after division by 2016 is 1 if and only if 2016 divides $x^2 - 1 = (x-1)(x+1)$. Let d be a common divisor of $x - 1$ and $x + 1$. Then d divides $(x+1) - (x-1) = 2$, and therefore $d \leq 2$. This means that at most one of the numbers $x-1$ and $x+1$ is divisible by 3 or by 7. If one of the numbers $x - 1$ and $x + 1$ is divisible by 2^k, where $k \geq 2$, the second is divisible by 2 only to the power 1. In order for 2^5 to divide $(x-1)(x+1)$, it is therefore necessary and sufficient that 2^4 divides one of the factors.

In order for $2016|(x-1)(x+1)$ to hold, it is therefore necessary and sufficient for some of the three numbers $\{2^4, 3^2, 7\}$ to divide $(x - 1)$ and the others to divide $(x+1)$. This yields $2^3 = 8$ combinations altogether. By the Chinese Remainder Theorem, each of the these possible combinations gives a unique solution modulo $2^4 \cdot 3^2 \cdot 1008$, and therefore exactly two solutions in the set $\{1, 2, \ldots, 2016\}$.

We see that there are $8 \cdot 2 = 16$ numbers in the set $\{1, 2, \ldots, 2016\}$ leaving a remainder of 1 after dividing their square by 2016.

Remark. These numbers are 1, 127, 433, 449, 559, 575, 881, 1007, 1009, 1135, 1141, 1457, 1567, 1583, 1889 and 2015.

Problem 22. (B-T-3-15) *Determine the number of all six-digit palindromes which are divisible by seven.*

Solution: A six-digit palindrome is a positive integer which is written in the form $\overline{abccba}$ with decimal digits $a \neq 0, b, c$. We have

$$\begin{aligned}\overline{abccba} &= 100\,001a + 10\,010b + 1\,100c \\ &= 7(14\,286a + 1\,430b + 157c) - (a - c).\end{aligned}$$

Such a number is divisible by 7 if and only if $(a - c)$ is divisible by 7. Because $a \neq 0$ and c are digits, we have $-8 \leq a - c \leq 9$. It therefore follows that $(a - c) \in \{-7, 0, 7\}$ must hold.

For $a - c = -7$ we have

$$(a, c) \in \{(1, 8), (2, 9)\},$$

for $a - c = 0$ we have

$$(a, c) \in \{(1, 1), (2, 2), (3, 3), (4, 4), (5, 5), (6, 6), (7, 7), (8, 8), (9, 9)\},$$

and finally for $a - c = 7$ we have

$$(a, c) \in \{(7, 0), (8, 1), (9, 2)\}.$$

This yields a total of 14 possibilities for (a, c).

In all cases, b is an arbitrary digit, and we see that there exist a total of $14 \cdot 10 = 140$ six-digit palindromes which are divisible by 7. □

Problem 23. (A-T-3-14) *We are given seven distinct positive integers. Prove that four of these can be chosen, such that their sum is divisible by four.*

Solution: After division by 4, each of the seven chosen numbers leaves a remainder from the set $\{0, 1, 2, 3\}$. Let us now consider three possible cases:

(a) If at least five of the chosen numbers have remainders 0 or 2, then we consider two subcases:

(a1): If at least four of these five numbers have the same remainder, we choose these four numbers and their sum is divisible by 4.

(a2): Otherwise at least two of these numbers leave the remainder 0 and simultaneously at least two of them leave the remainder 2. In this subcase, we take two numbers with the remainder 0 and two numbers with the remainder 2. Altogether, the sum of these four numbers is divisible by 4.

(b) If (exactly) four of the seven chosen numbers have remainder 0 or 2, then we will consider the following subcases:

(b1): If all of them have the same remainder, we take these four numbers and their sum is divisible by 4.

(b2): Otherwise there is at least one number with the remainder 2 and at least one with the remainder 0. We can then focus on the three other numbers with odd remainders. At least two of them have the same remainder. In this case we take two of these, one number with the remainder 0 and one number with the remainder 2. Such four numbers certainly have a sum divisible by 4.

(c) If at most three of the seven chosen numbers have remainders 0 or 2, we add 1 to all seven numbers. Now by (a) or (b) we can choose four of them with their sum divisible by 4. If we now subtract 1 from each of these numbers, their sum remains divisible by 4. This completes the proof. □

Remark. It is easy to check that this statement is generally not valid for six chosen numbers. For example, we can consider arbitrary numbers with the remainders 0, 0, 0, 1, 1, 1, respectively.

Problem 24. (B-T-3-12) *Determine all triples (x, y, z) of positive integers for which each of the three numbers x, y, z is a divisor of the sum $x+y+z$.*

Solution: Without loss of generality, we assume $x \leq y \leq z$. Since $z|(x + y + z)$, we have $z|(x + y)$ and therefore $z \leq x + y \leq 2z$, and so either $z = x + y$ or $2z = x + y$ must hold.

If $z = x + y$, we have $y|(x + z) = 2x + y$ and therefore $y|2x$. This implies either $y = x$ or $y = 2x$. The first case yields the two solutions $(x, x, 2x)$, and $(x, 2x, 3x)$.

If $2z = x + y$, the assumption $x \leq y \leq z$ yields $x = y = z$.

It is now easily checked that each of these triples indeed satisfies the required condition, and we are finished.

Conclusion. There are 10 possible triples satisfying the conditions of the problem. These are:

$$(x, x, x), (x, x, 2x), (x, 2x, x), (2x, x, x), (x, 2x, 3x), (x, 3x, 2x),$$

$$(2x, x, 3x), (2x, 3x, x), (3x, 2x, x), (3x, x, 2x),$$

where x is an arbitrary positive integer. □

Problem 25. (A-T-3-11) *Peter throws two dice simultaneously and then writes the number of dots showing on a blackboard. Find the smallest number k with the following property: After k throws Peter can always choose some of the written numbers, such that their product leaves the remainder* 1 *after division by* 13.

Solution: After each throw Peter writes down some number from the set $\{2, 3, 4, \ldots, 12\}$. If the first n throws all yield the number 2, the remainders of the product of all n twos on the blackboard after division by 13 are shown in the following table:

n	1	2	3	4	5	6	7	8	9	10	11	12
$2^n \pmod{13}$	2	4	8	3	6	12	11	9	5	10	7	1

In this case Peter needs at least 12 throws. We will now show that 12 throws are sufficient, no matter which numbers are written.

Let a_i be the number Peter writes on the blackboard after the ith throw. Let us define numbers s_i by

$$s_1 = a_1, \ s_2 = a_1a_2, \ s_3 = a_1a_2a_3, \ \ldots, \ s_{12} = a_1a_2 \cdot \ldots \cdot a_{12}.$$

Since none of the numbers a_1 is divisible by 13, the remainders of s_i after division by 13 are all from the set $\{1, 2, 3, \ldots, 12\}$. If there exists an index i such that the remainder of s_i after division by 13 is 1, we are finished. If not, we note that the 12 numbers $s_1, s_2, \ldots, s_{12}$ all have remainders in the set $\{2, 3, 4, \ldots, 12\}$, which contains only 11 elements. By the pigeonhole principle, there exist two indices i, j $(1 \leq i < j \leq 12)$ such that the numbers s_i and s_j have the same remainder after division by 13. This means that their difference is divisible by 13. We can now write

$$\begin{aligned} s_i - s_j &= a_1a_2 \ldots a_i - a_1a_2 \ldots a_j \\ &= a_1a_2 \ldots a_i(a_{i+1} \ldots a_j - 1) = s_i(a_{i+1} \ldots a_j - 1). \end{aligned}$$

Since s_i is not divisible by 13, the expression in brackets must be, and we see that $a_{i+1} \ldots a_j$ leaves the remainder 1 after division by 13. We can also note that the product $a_{i+1} \ldots a_j$ has at least two factors, since none of a_i by itself leaves the remainder 1, and we are finished.

We see that the smallest possible number k of throws is indeed 12, as claimed. □

Problem 26. (B-T-3-10) *How many positive integers of the form* $\overline{abcabcabc}$ *exist that are divisible by* 29?

Solution: We can write the number $\overline{abcabcabc}$ in the form

$$\overline{abcabcabc} = \overline{abc} \cdot 1\,001\,001.$$

Because 29 is not a divisor of 1 001 001, it follows that $29|\overline{abc}$.
The smallest three-digit number divisible by 29 is $4 \cdot 29 = 116$, and the largest is $34 \cdot 29 = 986$. Altogether, there are 31 numbers satisfying the required conditions. □

Problem 27. (B-I-1-10) *Prove that the number* $2010^{2011} - 2010$ *is divisible by* $2010^2 + 2011$.

Solution: We first note that

$$x^{2010} - 1 = (x^3 - 1)(x^{2007} + x^{2004} + \cdots + 1)$$

and

$$x^3 - 1 = (x - 1)(x^2 + x + 1)$$

both hold. We therefore have

$$(x^2 + x + 1)|(x^3 - 1) \quad \text{and} \quad (x^3 - 1)|(x^{2010} - 1),$$

and therefore also

$$(x^2 + x + 1)|(x^{2010} - 1).$$

Setting $x = 2010$ therefore yields

$$(2010^2 + 2010 + 1)|(2010^{2010} - 1),$$

and then also

$$(2010^2 + 2011)|[2010 \cdot (2010^{2010} - 1)] = 2010^{2011} - 2010,$$

as claimed. □

Problem 28. (B-T-1-09) *Determine the largest positive integer k such that the number* $n^6 - n^4 - n^2 + 1$ *is divisible by* 2^k *for every odd integer* $n > 1$.

Solution: After factorisation we have

$$n^6 - n^4 - n^2 + 1 = (n^2 - 1)^2(n^2 + 1).$$

Because n is an odd integer, we can write $n = 2k - 1$ for some positive integer k. We then obtain

$$(n^2-1)^2(n^2+1) = (4k^2-4k)^2(4k^2-4k+2) = 2^5(k-1)^2k^2(2k^2-2k+1).$$

The last factor in this expression is an odd integer. The factors $(k-1)^2$ and k^2 are the squares of two further integers of which one is odd and the other even. The product $(k-1)^2k^2$ is therefore divisible by the factor 2^2. We see that the number $n^6 - n^4 - n^2 + 1$ is divisible by 2^7 for every odd positive integer n.

Substituting $n = 3$, we obtain the number $3^6-3^4-3^2+1 = 640 = 2^7\cdot 5$, and we see that 7 is indeed the largest possible value of k. □

Problem 29. (B-T-1-03) *Determine all three-digit numbers $\overline{xyz}$ which are divisible by all three digits x, y, z and with no two digits being equal.*

Solution: We first note that $xyz \neq 0$, because we cannot divide the number $\overline{xyz}$ by 0.

Let us consider the last digit z of the number $\overline{xyz}$. The following cases are possible:

(a) z is odd. If the number $\overline{xyz}$ is divisible by x, y and z, both digits x and y must be also be odd. The total number of possibilities for $\overline{xyz}$ in this case is therefore equal to $5 \cdot 4 \cdot 3 = 60$. We can now check which of these numbers indeed fulfil the condition.

None of the numbers with $z = 1$, namely 351, 371, 391, 531, 571, 591, 731, 751, 791, 931, 951 and 971, is a solution.

None of the numbers with $z = 3$, namely 153, 173, 193, 513, 573, 593, 713, 753, 793, 913, 973 and 395, is a solution.

Of the numbers with $z = 5$, only 135, 175, 315 and 735 are solutions (and 195, 375, 395, 715, 795, 915, 935, 975 are not).

None of the numbers with $x = 7$, namely 137, 157, 197, 317, 357, 397, 517, 537, 597, 917, 937 and 957, is a solution.

None of the numbers with $z = 9$ can be a solution; by the divisibility rule for 9, the sum $x + y + z$, and therefore also $x + y$, would have to be divisible by 9, which is not possible for $x, y \in \{1, 3, 5, 7\}$.

We see that case (a) only yields four solutions.

(b) z is even. If the number $\overline{xyz}$ is divisible by x, y and z, none of the digits can be equal to 5, because the last digit would then be either 0 or 5. The last digit can only be 2, 4, 6 or 8. The total number of possibilities for $\overline{xyz}$ in this case is therefore equal to $4 \cdot 7 \cdot 6 = 168$. Once again, we can check which of these numbers indeed fulfil the condition.

Of the numbers with $z = 2$, the seven numbers 132, 162, 312, 412, 432, 612 and 672 are solutions (and 142, 172, 182, 192, 342, 362, 372, 382, 392, 462, 472, 482, 492, 632, 642, 682, 692, 712, 732, 742, 762, 782, 792, 812, 832, 842, 862, 872, 892, 912, 932, 942, 962 and 982 are not).

Of the numbers with $z = 4$, the eight numbers 124, 184, 264, 324, 384, 624, 784, 824 and 864 are solutions (and 134, 164, 174, 194, 214, 234, 274, 284, 294, 314, 364, 374, 394, 614, 634, 674, 684, 694, 714, 724, 734, 764, 794, 814, 834, 874, 894, 914, 924, 934, 964, 974 and 984 are not).

Of the numbers with $z = 6$, the five numbers 126, 216, 396, 816 and 936 are solutions (and 136, 146, 176, 186, 196, 236, 246, 276, 286, 296, 316, 326, 346, 376, 386, 416, 426, 436, 476, 486, 496, 716, 726, 736, 746, 786, 796, 826, 836, 846, 876, 896, 916, 926, 946, 976 and 986 are not).

Of the numbers with $z = 8$ the five numbers 128, 168, 248, 648 and 728 are solutions (and 138, 148, 178, 198, 218, 238, 268, 278, 298, 318, 328, 348, 368, 378, 398, 418, 428, 438, 468, 478, 498, 618, 628, 638, 678, 698, 718, 738, 748, 768, 798, 918, 928, 938, 948, 968 and 978 are not).

We see that case (b) yields $7 + 9 + 5 + 5 = 26$ solutions. Altogether, we therefore have $4 + 26 = 30$ solutions. □

Another solution: We note that $xyz \neq 0$ must hold, because we can certainly not divide the number $\overline{xyz}$ by 0. For this solution, we shall consider certain three-digit numbers that are divisible by 7.

The numbers $98a$ and $7b$ (where a and b are digits) are certainly divisible by 7. If we subtract the numbers $98a$ and $7b$ from a three-digit number $\overline{abc}$, we obtain

$$\overline{abc} - 98a - 7b = 2a + 3b + c.$$

A three-digit number $\overline{abc}$ is therefore divisible by 7 if and only if $2a+3b+c$ is divisible by 7.

We now begin our proof by assuming that a digit x, y, z of the number $\overline{xyz}$ is equal to 5. The digits are different, and since no digit can be 0, the divisibility rule for 5 tells us that this must be the last digit, yielding $z = 5$. By the divisibility rule for 2, none of the digits can be even. If a digit is equal to 7, the rule we have just developed shows us that only the numbers 735 and 175 are solutions. If a digit is equal to 3 (and no digit equal to 7), then the divisibility rule for 3 shows us that 135 and 315 are the only possible solutions. The remaining numbers 195 and 915 not yet considered are not solutions, as we easily check, giving us a total of four solutions in this case.

We can now assume that $x, y, z \in \{1, 2, 3, 4, 6, 7, 8, 9\}$ holds.

Assuming that one of the digits is equal to 9, the divisibility rule for 9 tells us that the sum of the other two digits is divisible by 9, and they can therefore only be the pairs of digits $\{1, 8\}$, $\{2, 7\}$ or $\{3, 6\}$. Of the numbers 918, 198, 972, 792, 936 and 396, only the 396 and 936 are solutions.

For the rest of the argument, we can assume that $x, y, z \in \{1, 2, 3, 4, 6, 7, 8\}$ holds.

Assuming now that one of the digits is equal to 7, we have three cases to consider:

Case 1: $x = 7$. The two-digit number $\overline{yz}$ with $y, z \in \{1, 2, 3, 4, 6, 8\}$ must be divisible by 7. Using the rule for divisibility by 7, we see that only the numbers 728 and 784 out of the candidates 714, 721, 728, 742, 763 and 784 are solutions.

Case 2: $y = 7$. Of the possible numbers 273, 371, 378, 476, 672 and 784, only 672 is a solution.

Case 3: $z = 7$. The number is not divisible by 2, and the digits x, y belong to the set $\{1, 3\}$. However, the numbers 137 and 317 are not solutions, because they are not divisible by 3.

These cases therefore yield the next three solutions, and we can now limit our considerations to $x, y, z \in \{1, 2, 3, 4, 6, 8\}$.

In this case, one digit must be even, as there are only two odd digits to choose from. The last digit must therefore be even. Furthermore, 3 and 6 cannot simultaneously be digits of the number. By the divisibility rule for 3, the third digit would also have to be 3, but there is no third such digit. If one of the digits is 3 or 6, the other two digits must therefore belong to one of the sets $\{1, 2\}$, $\{1, 8\}$, $\{2, 4\}$ or $\{3, 8\}$.

If the digit 3 belongs to the number, candidates are the numbers 132, 312, 138, 318, 432, 342, 234, 324, 834, 384, 438 and 348, of which 132, 312, 432, 324 are 384 solutions.

If the digit 6 belongs to the number (but $z \neq 6$), candidates are the numbers 162, 612, 168, 618, 462, 642, 264, 624, 864, 684, 468 are 648, of which 162, 612, 168, 264, 624, 864 and 648 are solutions.

If $z = 6$, candidates are the numbers 126, 216, 186, 816, 426, 246, 846 and 486, of which 126, 216 and 816 are solutions.

These cases therefore yield a total of $5 + 7 + 3 = 15$ further solutions.

Finally, we can now limit our discussion to $x, y, z \in \{1, 2, 4, 8\}$. Since $z \neq 1$, candidates of this type are the numbers 142, 412, 182, 812, 482, 842, 124, 214, 184, 814, 284, 824, 128, 218, 148, 418, 248 and 428, of which 412, 124, 184, 128, 248 and 824 are solutions, yielding a final five solutions.

Summing up, we see that the problem has exactly $4+2+3+15+6 = 30$ solutions. □

Problem 30. (A-I-2-02) *Prove that the expression* $2^{12n+8} - 3^{6n+2}$ *is divisible by* 13 *for every non-negative integer value* n.

Solution: We can prove this by induction.

For $n = 0$ we have $2^8 - 3^2 = 13 \cdot 19$.

We now assume that $13 | 2^{12n+8} - 3^{6n+2}$ holds for some positive integer n, i.e., there exists a positive integer k such that $2^{12n+8} - 3^{6n+2} = 13 \cdot k$. We now wish to show that this implies $13 | 2^{12(n+1)+8} - 3^{6(n+1)+2}$. This immediately follows from

$$\begin{aligned} 2^{12(n+1)+8} - 3^{6(n+1)+2} &= 2^{12} \cdot 2^{12n+8} - 3^6 \cdot 3^{6n+2} \\ &= 2^{12} \cdot (2^{12n+8} - 3^{6n+2}) + 3367 \cdot 3^{6n+2} \\ &= 2^{12} \cdot 13k + 13 \cdot 259 \cdot 3^{6n+2} \\ &= 13 \cdot (k \cdot 2^{12} + 3^{6n+2} \cdot 259), \end{aligned}$$

completing the proof. □

Another solution: In this proof, we consider congruences modulo 13. We first note that $2^4 \equiv 3 \pmod{13}$, $2^8 \equiv 9 \pmod{13}$ and $2^{12} \equiv 1 \pmod{13}$ hold. We therefore have $2^{12n} = (2^{12})^n \equiv 1 \pmod{13}$, and

$$2^{12n+8} = 2^8 \cdot 2^{12n} \equiv 9 \pmod{13}.$$

Similarly, we have $3^2 = 9 \equiv 9 \pmod{13}$ and $3^6 \equiv 1 \pmod{13}$. This yields $3^{6n} = (3^6)^n \equiv 1 \pmod{13}$, and we therefore have

$$3^{6n+2} = 3^{6n} \cdot 3^2 \equiv 9 \pmod{13}.$$

Subtraction therefore gives us

$$2^{12n+8} - 3^{6n+2} \equiv 0 \pmod{13},$$

completing the proof. □

Problem 31. (A-I-1-01) *Prove that*

$$L = m^{21}n^3 - m^7n - m^3n^{21} + mn^7$$

is divisible by 42 *for all integer values of m and n.*

Solution: We first show that L is divisible by 7. By Fermat's theorem, it follows that $a^7 \equiv a \pmod 7$ holds for an arbitrary integer a. From this, we obtain

$$L \equiv m^3n^3 - mn - m^3n^3 + mn = 0 \pmod 7.$$

Similarly, $a^3 \equiv a \pmod 3$ holds for an arbitrary integer a, and we have

$$L \equiv m^7n - mn - mn^7 + mn = 0 \pmod 3.$$

Finally, L is certainly divisible by 2, since the four terms of L are either all even or all odd, and L thus certainly even.

Summarising, we see that L is divisible by the product of the prime numbers 2, 3 and 7, and since $2 \cdot 3 \cdot 7 = 42$ holds, this completes the proof. □

Problem 32. (B-I-4-99) *Prove that the number* $m^5n - mn^5$ *is divisible by* 30 *for all positive integers m and n.*

Solution: Of course, this problem is closely related to the previous one. Still, there are some interesting differences. We define $L = m^5n - mn^5$ and rewrite in the form

$$L = m^5n - mn^5 = mn(m-n)(m+n)(m^2+n^2).$$

To prove that L is divisible by 30, we shall prove that it is simultaneously divisible by 2, 3 and 5.

First of all, L is certainly divisible by 2. If either of the two numbers m, n is even, L is obviously even. If m and n are both odd, then $m + n$ is even.

Either way, we certainly have $2|L$.

Next, we prove that $3|L$. If either of the numbers m, n is divisible by 3, we certainly have $3|L$. If neither of the numbers m, n is divisible by 3, each of them gives the remainder 1 or 2 modulo 3. We can therefore write $m = 3k + p$ and $n = 3l + q$, where k, l are non-negative integers and $p, q \in \{1, 2\}$. If $p = q$, we have

$$m - n = 3k + p - (3l + q) = 3k + p - 3l - q = 3(k - l),$$

and the factor $n - m$ is divisible by 3. On the other hand, if $p \neq q$, $p, q \in \{1, 2\}$ we have

$$n + m = 3k + p + 3l + q = 3k + 3l + 1 + 2 = 3(k + l + 1),$$

and the factor $m + n$ is divisible by 3. In both cases, we have $3|L$.

Finally, we prove that $5|L$. If either of the numbers m, n is divisible by 5, we certainly have $5|L$. If neither of the numbers m, n is divisible by 5, each of them gives the remainder 1, 2, 3 or 4 modulo 5. We can therefore write $m = 5k + p$ and $n = 5l + q$, where k, l are non-negative integers and $p, q \in \{1, 2, 3, 4\}$. If $p = q$, we have

$$m - n = 5k + p - (5l + q) = 5k + p - 5l - q = 5(k - l),$$

and the factor $n - m$ is divisible by 5. If we have $p \neq q$ with $p, q \in \{1, 2, 3, 4\}$, we can consider all possible pairs as follows:

- if $p, q \in \{1, 2\}$, then $m^2 + n^2 \equiv 5 \pmod 5$;
- if $p, q \in \{1, 3\}$, then $m^2 + n^2 \equiv 10 \pmod 5$;
- if $p, q \in \{1, 4\}$, then $m + n \equiv 5 \pmod 5$;
- if $p, q \in \{2, 3\}$, then $m + n \equiv 5 \pmod 5$;
- if $p, q \in \{2, 4\}$, then $m^2 + n^2 \equiv 20 \pmod 5$;
- if $p, q \in \{3, 4\}$, then $m^2 + n^2 \equiv 25 \pmod 5$.

In all of these cases, we have $5|L$, and this completes the proof. □

Problem 33. (B-I-4-98) *Prove that a two-digit number, which is not divisible by* 10 *but is divisible by the sum of its digits, is certainly divisible by* 3. *Is this also true for all three-digit numbers with this property*?

Solution: Let $\overline{ab}$ be a two-digit number of this type. The number is not divisible by 10, and we therefore have $b \neq 0$.

From $a + b \mid 10a + b$, we obtain $a + b \mid 10(a + b) - (10a + b)$, or $a + b \mid 9b$.

If $a + b$ is not divisible by 3, this implies $a + b \mid b$, which is not possible because the digits a and b are positive.

We therefore certainly have $3 \mid a + b$, and since $3 \mid 9a$ obviously holds, we obtain $3 \mid (a + b) + 9a = 10a + b$, completing the proof.

The theorem is not true for three-digit numbers. A counterexample is the number 322. (We immediately see that $7 \mid 322$ holds, but 322 is not divisible by 3.) □

Problem 34. (B-I-3-96) *We are given the set of all seven-digit numbers, in which the digits are some permutation of* $1, 2, 3, \ldots, 7$. *Prove that the sum of all these numbers is divisible by* 9.

Solution: Let $d \in \{1, 2, 3, \ldots, 7\}$ be the units digit of some number in the set. There are 6! numbers in the set with this units digit. The sum of all these digits is therefore $d \cdot 6!$, and the sum of the units digits of all numbers of the set is therefore equal to $6!(1 + 2 + 3 + 4 + 5 + 6 + 7) = 6! \cdot 28$.

If we now take a look at the tens digits, the hundreds digits, the thousands digits and so on, we similarly obtain the sums $10 \cdot 6! \cdot 28$, $10^2 \cdot 6! \cdot 28$, $10^3 \cdot 6! \cdot 28, \ldots, 10^6 \cdot 6! \cdot 28$, respectively. Adding all of these sums, we obtain

$$6! \cdot 28 \cdot (1 + 10 + 10^2 + \cdots + 10^6) = 6! \cdot 28 \cdot 1111111.$$

The number 9 certainly divides $6! = 6 \cdot 5 \cdot 4 \cdot 4 \cdot 3 \cdot 2 \cdot 1$, and therefore also $6! \cdot 28 \cdot 1111111$, as claimed. □

Problem 35. (A-I-1-93) *Prove that the number*

$$1993^2 - 1992^2 + 1991^2 - 1990^2 + \cdots + 3^2 - 2^2 + 1^2$$

is divisible by 1993.

Solution: We can rewrite the given number in the form

$$\begin{aligned}&(1993 - 1992) \cdot (1993 + 1992) + (1991 - 1990) \cdot (1991 + 1990)\\&\quad + \cdots + (3 - 2) \cdot (3 + 2) + 1\\&= 1993 + 1992 + 1991 + 1990 + \cdots + 3 + 2 + 1.\end{aligned}$$

This is an arithmetic series, and the sum is therefore equal to

$$\frac{1993+1}{2}\cdot 1993 = 997\cdot 1993.$$

This is certainly divisible by 1993, and we are finished. □

Remark. For every odd positive integer n, the number

$$n^2-(n-1)^2+(n-2)^2-(n-3)^2+\cdots+3^2-2^2+1^2$$

is certainly divisible by n.

2.4. Diophantine Equations

The subject of integer-valued solutions to equations of various types is a very popular one in the Mathematical Duel. Solving these problems can require a quite diverse combination of tools from algebra and number theory, always combined with a bit of clever insight, of course.

PROBLEMS

36. The positive integers k, l, m, n fulfil the equation

$$k^2l^2-m^2n^2 = 2015 + l^2m^2 - k^2n^2.$$

Find all possible values of $k+l+m+n$.

37. Determine all pairs (x, y) of integers fulfilling the equation

$$x^2-3x-4xy-2y+4y^2+4=0.$$

38. How many triples (a, b, c) of positive integers with

$$abc = 45000$$

exist?

39. Determine all integer solutions of the equation

$$x^6 = y^3 + 4069.$$

40. Determine all integer solutions (x, y) of the equation

$$\frac{2}{x}+\frac{3}{y}=1.$$

41. Solve the following equation in positive integers

$$xyz = 2x + 3y + 5z.$$

42. Determine all pairs (p, x) fulfilling the following equation:

$$x^2 = p^3 + 1,$$

where p is a prime and x is an integer.

43. Prove that there exist infinitely many solutions of the equation

$$2^x + 2^{x+3} = y^2$$

in the domain of positive integers.

44. Solve the following equation in the domain of positive integers:

$$\frac{2}{x^2} + \frac{3}{xy} + \frac{4}{y^2} = 1.$$

45. Determine all pairs (x, y) of integers such that the equation

$$4^x = 1899 + y^3$$

is fulfilled.

46. Determine all pairs (x, y) of positive integers such that the equation

$$4^x = y^2 + 7$$

is fulfilled.

47. Determine all triples (x, y, z) of positive integers such that

$$3 + x + y + z = xyz$$

holds.

48. Determine all integers $a > b > c > d > e$ such that

$$(5 - a)(5 - b)(5 - c)(5 - d)(5 - e) = 20.$$

49. Determine all pairs (x, y) of positive integers such that

$$17x^2 + 1 = 9y! + 2004.$$

50. Solve the following equation in integers

$$x^2 + y^2 = xy + 4x + 4y.$$

51. Determine all integer solutions of the equation

$$(x^2 - y^2)^2 = 16y + 1.$$

52. Determine all pairs (x, y) of integers fulfilling the equation

$$x^3 - 4x^2 - 5x = 6^y.$$

53. Determine all triples (x, y, z) of positive integers fulfilling the equation

$$\frac{1}{x} + \frac{2}{y} + \frac{3}{z} = 4.$$

54. Determine all integer solutions of the equation

$$xyz + xy + yz + xz + x + y + z = 1996.$$

55. Determine as many triples of different positive integers (k, l, m) as possible, fulfilling the equation

$$\frac{1}{1995} = \frac{1}{k} + \frac{1}{l} + \frac{1}{m}.$$

56. Determine all pairs (x, y) of integers fulfilling the equation

$$\sqrt{x} + \sqrt{y} = \sqrt{90}.$$

SOLUTIONS

Problem 36. (A-I-4-15) *The positive integers k, l, m, n fulfil the equation*

$$k^2l^2 - m^2n^2 = 2015 + l^2m^2 - k^2n^2.$$

Find all possible values of $k + l + m + n$.

Solution: The given equation can be rewritten as

$$(k^2 - m^2)(l^2 + n^2) = 2015.$$

We have $2015 = 5 \cdot 13 \cdot 31$ with primes 5, 13 and 31 and

$$5 \equiv 1 \pmod 4), \quad 13 \equiv 1 \pmod 4), \quad 31 \equiv 3 \pmod 4).$$

Since the sum of two square numbers is never congruent 3 modulo 4, and $l^2 + n^2 > 1$ holds, it follows that $l^2 + n^2$ is not a multiple of 31, and

$$l^2 + n^2 \in \{5, 13, 65\}.$$

We now consider three possible cases:

(a) $l^2+n^2 = 5$. The only possible representation of 5 as a sum of two square numbers is $5 = 1^2 + 2^2$. It follows immediately that $\{l, n\} = \{1, 2\}$ and $l + n = 3$. This means that

$$k^2 - m^2 = (k + m)(k - m) = 31 \cdot 13.$$

Since $k + m > k - m$ we get

$$k + m = 31, \quad k - m = 13 \quad \text{or} \quad k + m = 13 \cdot 31 = 403, \quad k - m = 1.$$

In both cases, k and m are positive integers, so $(l + n) + (k + m)$ can be equal to $3 + 31 = 34$ or equal to $3 + 403 = 406$.

(b) $l^2 + n^2 = 13$. It follows that $l^2 + n^2 = 3^2 + 2^2$, $l + n = 5$ and, consequently

$$k^2 - m^2 = (k + m)(k - m) = 5 \cdot 31.$$

This implies that

$$k + m = 31, \quad k - m = 5 \quad \text{or} \quad k + m = 5 \cdot 31 = 155, \quad k - m = 1.$$

Again, k and n are positive integers, so $(l + n) + (k + m)$ can be equal to $5 + 31 = 36$ or $5 + 155 = 160$.

(c) $l^2 + n^2 = 65$. We must deal with two subcases: $l^2 + n^2 = 8^2 + 1^2$ or $l^2 + n^2 = 7^2 + 4^2$, so we have $l + n = 8 + 1 = 9$ or $l + n = 7 + 4 = 11$. It follows that

$$k^2 - m^2 = (k + m)(k - m) = 31,$$

and $k + m = 31$, $k - m = 1$. Since k and n are positive integers, $(l + n) + (k + m)$ can attain the values $9 + 31 = 40$ and $11 + 31 = 42$.

Summing up cases (a), (b) and (c), we obtain

$$k + l + m + n \in \{34, 36, 40, 42, 160, 406\}.$$ □

Problem 37. (A-T-1-15) *Determine all pairs* (x, y) *of integers fulfilling the equation*

$$x^2 - 3x - 4xy - 2y + 4y^2 + 4 = 0.$$

Solution: We can rewrite the equation in the form

$$(x-2y)^2 = 3x + 2y - 4.$$

Since both x and y are integers, there exists an integer d such that

$$x - 2y = d,$$
$$3x + 2y - 4 = d^2.$$

Solving this system we obtain

$$x = \frac{d(d+1)}{4} + 1, \qquad y = \frac{d(d-3)+4}{8}.$$

We check all residues modulo 8 to obtain that x and y are integers if and only if $d = 8k + 4$ ($k \in \mathbb{Z}$) or $d = 8(k-1) + 7 = 8k - 1$ ($k \in \mathbb{Z}$). In the first case we have

$$x = 16k^2 + 18k + 6, \qquad y = 8k^2 + 5k + 1, \quad k \in \mathbb{Z}, \tag{1}$$

and the second case gives us

$$x = 16k^2 - 2k + 1, \qquad y = 8k^2 - 5k + 1, \quad k \in \mathbb{Z}, \tag{2}$$

and we see that all integer solutions of the given equation are in the form (1) or (2). □

Problem 38. (B-I-1-15) *How many triples* (a, b, c) *of positive integers with*

$$abc = 45\,000$$

exist?

Solution: Note that $45\,000 = 2^3 \cdot 3^2 \cdot 5^4$. Any positive integer solution (a, b, c) of the given equation must be of the form

$$a = 2^{\alpha_1} \cdot 3^{\beta_1} \cdot 5^{\gamma_1},$$
$$b = 2^{\alpha_2} \cdot 3^{\beta_2} \cdot 5^{\gamma_2},$$
$$c = 2^{\alpha_3} \cdot 3^{\beta_3} \cdot 5^{\gamma_3},$$

where exponents $\alpha_i, \beta_i, \gamma_i$ ($i = 1, 2, 3$) are non-negative integers fulfilling the following system of equations:

$$\alpha_1 + \alpha_2 + \alpha_3 = 3, \tag{1}$$

$$\beta_1 + \beta_2 + \beta_3 = 2, \tag{2}$$

$$\gamma_1 + \gamma_2 + \gamma_3 = 4. \tag{3}$$

Equation (1) is satisfied by triples $(\alpha_1, \alpha_2, \alpha_3)$ from the set

$$\{(3,0,0), (0,3,0), (0,0,3), (2,1,0), (1,2,0), (2,0,1), (1,0,2), (0,1,2), (0,2,1), (1,1,1)\},$$

i.e., 10 triples altogether.

Equation (2) is satisfied by triples $(\beta_1, \beta_2, \beta_3)$ from the set

$$\{(2,0,0), (0,2,0), (0,0,2), (1,1,0), (1,0,1), (0,1,1)\},$$

i.e., 6 triples altogether.

Equation (3) is satisfied by triples $(\gamma_1, \gamma_2, \gamma_3)$ from the set

$$\{(4,0,0), (0,4,0), (0,0,4), (3,1,0), (1,3,0), (3,0,1), (1,0,3), (0,1,3), (0,3,1), (2,2,0), (2,0,2), (0,2,2), (2,1,1), (1,2,1), (1,1,2)\}$$

i.e., 15 triples altogether.

By the combinatorial product principle, it follows that the number of all triples $(\alpha_1, \alpha_2, \alpha_3)$, $(\beta_1, \beta_2, \beta_3)$, $(\gamma_1, \gamma_2, \gamma_3)$ fulfilling (1), (2) and (3) is equal to $10 \cdot 6 \cdot 15 = 900$. □

Problem 39. (B-T-1-14) *Determine all integer solutions of the equation*

$$x^6 = y^3 + 4\,069.$$

Solution: We can rewrite the given equation in the equivalent form

$$(x^2 - y)(x^4 + x^2y + y^2) = 4\,096.$$

Since $x^4 + x^2y + y^2 = (x^2 + \frac{1}{2}y)^2 + \frac{3}{4}y^2 \geq 0$, both factors $x^2 - y$ and $x^4 + x^2y + y^2$ must be positive.

First of all, we can see that for each $y \in \{-2, -1, 0\}$ no integer x fulfilling the given equation exists. For these values of y, we have $x^6 \in \{4\,061, 4\,068, 4\,069\}$, which is not possible for integer values of x.

We now show that the inequality

$$x^2 - y < x^4 + x^2y + y^2$$

must hold for any other integer y. Rewriting this inequality, we obtain

$$x^4 + (y-1)x^2 + y^2 + y > 0$$

with unknown x. For the variable x^2 and an arbitrary integer parameter $y \notin \{-2, -1, 0\}$, the associated quadratic equation has a negative discriminant Δ, because

$$\Delta = (y-1)^2 - 4(y^2 + y) = 4 - 3(y+1)^2 < 0.$$

Since $0 < x^2 - y < x^4 + x^2y + y^2$ and 4069 is the product of two primes 13 and 313, we now have only two possible cases to consider:

(a) $x^2 - y = 1$ and $x^4 + x^2y + y^2 = 4\,069$. Substituting $x^2 = y + 1$ we obtain the equation $y^2 + y - 1\,356 = 0$ with the discriminant $\Delta = 5\sqrt{217}$ and y is therefore not an integer.

(b) $x^2 - y = 13$ and $x^4 + x^2y + y^2 = 313$. Substituting $x^2 = y + 13$ we obtain the equation $y^2 + 13y - 48 = 0$ with roots $y_1 = -16$ and $y_2 = 3$. For $y_1 = -16$ we have $x^2 = -3$, which is impossible. For $y_2 = 3$ we have $x^2 = 16$, and therefore $x_1 = -4$ and $x_2 = 4$.

We see that the given equation has two integer solutions, namely $(x, y) \in \{(-4; 3), (4; 3)\}$. □

Problem 40. (B-T-2-13) *Determine all integer solutions* (x, y) *of the equation*

$$\frac{2}{x} + \frac{3}{y} = 1.$$

Solution: We can rewrite the given equation in the form

$$xy - 3x - 2y + 6 = 6 \iff (x-2)(y-3) = 6.$$

The integer 6 can be factored as the product of two integers as follows:

$$6 = 1 \cdot 6 = 6 \cdot 1 = 2 \cdot 3 = 3 \cdot 2 = (-1) \cdot (-6)$$
$$= (-6) \cdot (-1) = (-2) \cdot (-3) = (-3) \cdot (-2).$$

Therefore, we have eight possible cases, which are collected in the following table:

$x-2$	1	6	2	3	-1	-6	-2	-3
$y-3$	6	1	3	2	-6	-1	-3	-2
x	3	8	4	5	1	-4	0	-1
y	9	4	6	5	-3	2	0	1

Since $x \neq 0$ and $y \neq 0$, the given equation has exactly seven solutions, namely

$$(x, y) \in \{(3; 9), (8; 4), (4; 6), (5; 5), (1; -3), (-4; 2), (-1, 1)\}. \quad \square$$

Problem 41. (A-T-1-12) *Solve the following equation in positive integers*

$$xyz = 2x + 3y + 5z.$$

Solution: First of all, we discuss some special cases for x and y.

For $x = 1$ we have to solve $yz = 2 + 3y + 5z$ which is equivalent to $(y-5)(z-3) = 17$. Since $y - 5 > -4$ and $z - 3 > -2$, both factors are positive integer factors of the prime 17, and so we have solutions

$$(x, y, z) \in \{(1, 6, 20), (1, 22, 4)\}.$$

For $x = 2$ we have the equation $2yz = 4 + 3y + 5z$ which is equivalent to $(2y-5)(2z-3) = 23$. In the same way, we obtain solutions

$$(x, y, z) \in \{(2, 3, 13), (2, 14, 2)\}.$$

For $y = 1$ we have to solve $xz = 2x + 3 + 5z$ which is equivalent to $(x-5)(z-2) = 13$. Further solutions are then

$$(x, y, z) \in \{(6, 1, 15), (18, 1, 3)\}.$$

For $y = 2$ we have the equation $2xz = 2x + 6 + 5z$ which is equivalent to $(2x - 5)(z - 1) = 11$. Thus, we have solutions

$$(x, y, z) \in \{(3, 2, 12), (8, 2, 2)\}.$$

Now we can assume $x, y \geq 3$. In this case, we have $x - \frac{5}{y} \geq \frac{4}{3}$ and $y - \frac{5}{x} \geq \frac{4}{3}$. From $xyz = 2x + 3y + 5z$ we then obtain

$$z = \frac{2x}{xy - 5} + \frac{3y}{xy - 5} = \frac{2}{y - \frac{5}{x}} + \frac{3}{x - \frac{5}{y}} \leq 2 \cdot \frac{3}{4} + 3 \cdot \frac{3}{4} = \frac{15}{4},$$

and therefore $z \leq 3$.

Finally, we have to discuss three possibilities for z.

(a) If $z = 1$ we must solve the equation $xy = 2x + 3y + 5$ which is equivalent to $(x - 3)(y - 2) = 11$. From this, we obtain two solutions

$$(x, y, z) \in \{(4, 13, 1), (14, 3, 1)\}.$$

(b) If $z = 2$ we must solve the equation $2xy = 2x + 3y + 10$ which is equivalent to $(2x - 3)(y - 1) = 13$. In this case we get

$$(x, y, z) \in \{(2, 14, 2), (8, 2, 2)\}.$$

(c) Finally, if $z = 3$ we must solve the equation $3xy = 2x + 3y + 15$ which is equivalent to $(x - 1)(3y - 2) = 17$. In this case we have only one solution in positive integers $(x, y, z) \in \{(18, 1, 3)\}$, which was already obtained previously.

In summary, we see that the set of all solutions of the original equation in positive integers is the set

$$\{(1, 6, 20), (1, 22, 4), (2, 3, 13), (2, 14, 2),$$
$$(3, 2, 12), (4, 13, 1), (6, 1, 15), (8, 2, 2,), (14, 13, 1), (18, 1, 3)\}.$$

□

Problem 42. (B-I-1-12) *Determine all pairs (p, x) fulfilling the following equation:*

$$x^2 = p^3 + 1,$$

where p is a prime and x is an integer.

Solution: First of all, we rewrite the given equation in the form

$$(x-1)(x+1) = p^3.$$

If (x, p) is a solution of the equation, then $(-x, p)$ is also a solution. Therefore, without loss of generality, we can assume $x \geq 0$. We have $x-1 < x+1$ and we have only two possibilities to consider:

(a) $x - 1 = 1$ and $x + 1 = p^3$. This implies $x = 2$ and $p^3 = 3$ and we therefore have no solution in this case.

(b) $x - 1 = p$ and $x + 1 = p^2$. Subtracting these two equations, we obtain the quadratic equation $p^2 - p - 2 = 0$, with two roots $p = -1$ (which does not fulfil the conditions of the problem) and $p = 2$. This then yields $x = 3$.

The given equation therefore has exactly two solutions, namely

$$(x, p) \in \{(3, 2), (-3, 2)\}.$$ □

Problem 43. (B-I-3-11) *Prove that there exist infinitely many solutions of the equation*

$$2^x + 2^{x+3} = y^2$$

in the domain of positive integers.

Solution: We rewrite the given equation in the following way:

$$2^x + 2^{x+3} = 2^x \cdot 3^2 = y^2.$$

If $x = 2n$, where n is a positive integer, we obtain $y = 3 \cdot 2^n$. Every pair (x, y) obtained in this manner is therefore a solution of given equation for any arbitrary positive integer n, and the given equation therefore has infinitely many solutions. □

Problem 44. (B-T-1-11) *Solve the following equation in the domain of positive integers:*

$$\frac{2}{x^2} + \frac{3}{xy} + \frac{4}{y^2} = 1.$$

Solution: We note that the inequalities $x \geq 2$ and $y \geq 3$ must hold for any solution, because we certainly have $\frac{2}{x^2} < 1$ and $\frac{4}{y^2} < 1$.

For $y = 3$, the given equation reduces to

$$\frac{2}{x^2} + \frac{1}{x} = \frac{5}{9}.$$

From this, we obtain the equation $5x^2 - 9x - 18 = 0$. The first solution $x_1 = -\frac{6}{5}$ of this is not positive integer, but the second $x_2 = 3$ is. This case therefore yields one solution, namely $x = y = 3$.

We will now prove that this equation does not have any other positive solutions.

For $y = 4$, we obtain the equation

$$\frac{2}{x^2} + \frac{3}{4x} + \frac{1}{4} = 1,$$

which can be written in the form $3x^2 - 3x - 8 = 0$. This equation has the discriminant $\Delta = 105$ and thus this equation has no solution in positive integers.

For $y \geq 5$, recalling that $x \geq 2$ holds, we have

$$\frac{2}{x^2} \leq \frac{25}{50}, \qquad \frac{3}{xy} \leq \frac{15}{50}, \qquad \frac{4}{y^2} \leq \frac{8}{50},$$

and therefore

$$\frac{2}{x^2} + \frac{3}{xy} + \frac{4}{y^2} \leq \frac{48}{50}.$$

We see that the equation has a unique solution in positive integers, namely $x = y = 3$.

Remark. The proof that no solutions other than $x = y = 3$ exist can also be done in another way. Assume that there exists some other solution of the equation. In this case, the value of one of the variables must be smaller than 3 and the value of the other must be greater than 3 and obtain a sum equal to 1.

From the inequality $y \geq 3$, we know that y cannot be smaller than 3. Therefore, x must be smaller than 3, and from the inequality $x \geq 2$, the case $x = 2$ remains as the sole possible option.

For $x = 2$, we have the equation

$$\frac{2}{2^2} + \frac{3}{2y} + \frac{4}{y^2} = 1.$$

We can rewrite this equation in the form $y^2 - 3y - 8 = 0$, and this equation has the discriminant $\Delta = 41$. Since this is not a perfect square, it follows that there are no further positive integer solutions. □

Problem 45. (A-T-3-10) *Determine all pairs* (x, y) *of integers such that the equation*

$$4^x = 1899 + y^3$$

is fulfilled.

Solution: We note that $x \geq 0$ certainly must hold for any integer solution. From

$$4^x \equiv 1 \pmod 9 \quad \text{for } x \equiv 0 \pmod 3,$$
$$4^x \equiv 4 \pmod 9 \quad \text{for } x \equiv 1 \pmod 3,$$
$$4^x \equiv 7 \pmod 9 \quad \text{for } x \equiv 2 \pmod 3,$$

and the relations

$$y^3 \equiv 0 \pmod 9 \quad \text{for } y \equiv 0 \pmod 3,$$
$$y^3 \equiv 1 \pmod 9 \quad \text{for } y \equiv 1 \pmod 3,$$
$$y^3 \equiv 8 \pmod 9 \quad \text{for } y \equiv 2 \pmod 3,$$

as well as

$$1899 \equiv 0 \pmod 9,$$

it follows that only

$$x \equiv 0 \pmod 3 \quad \text{and} \quad y \equiv 1 \pmod 3$$

can hold for any integer solutions. We can therefore write $x = 3n$, where n is an integer, $n \geq 0$.

Note that $1899 = 3^2 \cdot 211$ and 211 is a prime. We can therefore write the equation in the equivalent form

$$3^2 \cdot 211 = (4^n - y)(4^{2n} + 4^n \cdot y + y^2).$$

The expression

$$4^{2n} + 4^n \cdot y + y^2 = \left(4^n + \frac{1}{2}m\right)^2 + \frac{3}{4}m^2$$

takes only positive values, and so we certainly have $4^n - y > 0$. Moreover, we also have

$$4^n - y \equiv 1 - 1 \equiv 0 \pmod{3}$$

and

$$4^{2n} + 4^n \cdot y + y^2 \equiv 1 + 1 + 1 \equiv 0 \pmod{3}.$$

It therefore remains to solve two systems of equations, namely

$$4^n - y = 633$$
$$4^{2n} + 4^n \cdot y + y^2 = 3$$

and

$$4^n - y = 3,$$
$$4^{2n} + 4^n \cdot y + y^2 = 633.$$

It is easy to check that the first of these has no solution. Substituting $4^n = 3 + y$ in the second equation of the second system, we obtain a quadratic equation $3(y^2 + 3y + 3) = 633$ with the roots $y_1 = -16$ and $y_2 = 13$.

The equation $4^n = 3 + m$ has an integer solution only for $m = 13$, and this solution is $n = 2$, yielding $x = 3n = 6$, and therefore $(x, y) = (6, 13)$. □

Problem 46. (B-I-2-10) *Determine all pairs (x, y) of positive integers such that the equation*

$$4^x = y^2 + 7$$

is fulfilled.

Solution: The given equation is equivalent to

$$4^x - y^2 = 7$$

or

$$(2^x - y)(2^x + y) = 7.$$

Since $2^x + y$ is a positive integer and $2^x - y$ is an integer, it follows that $2^x + y = 7$ and $2^x - y = 1$ must hold. From the first equation we have

$y = 2^x - 1$. Substituting y in the given equation we get

$$4^x - (2^x - 1)^2 = 7,$$

and therefore

$$2^{x+1} = 8.$$

This yields the unique solution $x = 2$ and $y = 3$. □

Problem 47. (A-T-1-08) *Determine all triples* (x, y, z) *of positive integers such that*

$$3 + x + y + z = xyz$$

holds.

Solution: By symmetry, we can assume, without loss of generality, that $1 \le x \le y \le z$ holds.

If $x = 1$, the equation can be expressed as $y + 4 = z(y - 1)$ (which is incorrect for $y = 1$), and we therefore have

$$z(y) = \frac{y+4}{y-1} = 1 + \frac{5}{y-1} \quad \text{for } y > 1.$$

For $y = 2$ we get $z = 6$, and since $3+1+2+6 = 12$ is a true statement, we see that all permutations of the triple $(1, 2, 6)$ are solutions of the equation. We shall now show that there are no other solutions.

Staying with the case $x = 1$, we note that $z(3) = \frac{7}{2}$ is not an integer and $z(3) = \frac{8}{3}$ is smaller than $y = 4$. Since $z(y)$ is a decreasing function, the existence of any further solutions would contradict $y \le z$.

For $x = 2$ we obtain

$$z(y) = \frac{y+5}{2y-1} = \frac{1}{2} + \frac{7}{2y-1} \quad \text{for } y \ge 2.$$

We see that $z(2) = \frac{7}{2}$ is not an integer, and since $z(3) = \frac{8}{5} < 3$, any further solutions in this case would once more contradict $y \le z$.

For any given $x \ge 3$ the assumption $x \le y \le z$ gives us

$$3 + x + y + z \le 3 + 3z \le z + 3z < 3 \cdot 3 \cdot z \le xyz,$$

and in this case, there can also be no further solutions of the equation in positive integers.

The solutions of the given equation are therefore precisely all permutations of the triple $(1, 2, 6)$. □

Problem 48. (B-I-1-06) *Determine all integers $a > b > c > d > e$ such that*

$$(5-a)(5-b)(5-c)(5-d)(5-e) = 20.$$

Solution: Because the integers a, b, c, d, e fulfil the inequalities $a > b > c > d > e$, the factors $5-a, 5-b, 5-c, 5-d, 5-e$ must be mutually different. Since these are all divisors of the number 20, they must all belong to the set $\{\pm 20, \pm 10, \pm 5, \pm 4, \pm 2, \pm 1\}$.

We now first note that neither 20 nor -20 can be one of these factors, because the other four factors would then all have to be equal to ± 1, and could therefore not all be different.

Similarly, neither 10 nor -10 can be one of these factors, because one of the other four factors would then be either 2 or -2, and all three of the others equal to ± 1. Once again, not all can be different.

If one of the factors were equal to -5, the product of the other four different factors from the set $\{\pm 1, \pm 2, \pm 4\}$ would have to be equal to -4, which is impossible.

If one of the factors is 5, then the product of the other four different factors from the set $\{\pm 1, \pm 2, \pm 4\}$ is equal to 4. The only way this can be possible is the case $4 = 1 \cdot (-1) \cdot 2 \cdot (-2)$.

Since the factor 5 must be contained in one of the factors, there are no other possible cases. We see that the distribution $20 = 5 \cdot 1 \cdot (-1) \cdot 2 \cdot (-2)$ fulfilling the assumptions is unique.

Subtracting the numbers $5, -2, 2, -1, 1$ from 5 gives us the numbers $0, 7, 3, 6, 4$. Because of the assumption $a > b > c > d > e$, we therefore have $a = 7$, $b = 6$, $c = 4$, $d = 3$ and $e = 0$.

Another solution: Because of the given properties of the variables, we certainly have

$$5-a < 5-b < 5-c < 5-d < 5-e.$$

All five different factors cannot be positive integers, because $5! = 120 > 20$. The product must therefore contain either two or four negative integer factors. Because $20 = 2 \cdot 2 \cdot 5$, it can be only the product $(-2) \cdot (-1) \cdot 1 \cdot 2 \cdot 5$. We therefore obtain $5-a = -2$, $5-b = -1$, $5-c = 1$, $5-d = 2$ and $5-e = 5$, yielding the unique solution $a = 7$, $b = 6$, $c = 4$, $d = 3$ and $e = 0$. □

Problem 49. (A-I-4-04) *Determine all pairs* (x, y) *of positive integers such that*

$$17x^2 + 1 = 9y! + 2004.$$

Solution: We solve the equivalent equation

$$17x^2 = 9y! + 2003.$$

For $y = 1$, we have the equation $17x^2 = 2012$. This equation has no integer solution. Similarly, for $y = 2$ we obtain the equation $17x^2 = 2021$, which also has no integer solution.

For $y = 3$ we have the equation $17x^2 = 2057$, which gives us $x^2 = 121$, and we therefore obtain the solution $x = 11$, $y = 3$.

For $y \in \{4, 5, \ldots, 16\}$, we can write the equivalent equations as above, none of which has an integer solution.

Finally, we note that for $y \geq 17$ we have $17 | 17x^2$ and $17 | 9y!$. However, 17 is not a divisor of the number 2003. It therefore follows that the equation has no further solutions for $y \geq 17$, and we see that the unique solution of the given equation is $x = 11$, $y = 3$. □

Problem 50. (A-T-2-00) *Solve the following equation in integers*

$$x^2 + y^2 = xy + 4x + 4y.$$

Solution: If (x, y) is a integer solution with $y \geq x$, we define $m = y - x \geq 0$. We then have $y = x + m$, and substitution in the given equation gives us

$$x^2 + (x + m)^2 = x(x + m) + 4x = 4(x + m).$$

This is equivalent to the quadratic equation

$$x^2 + (m - 8)x + m^2 - 4m = 0 \tag{1}$$

with the discriminant $\Delta = -3m^2 + 64$. If a solution of (1) exists, we certainly have $\Delta \geq 0$ and therefore $|m| \in \{0, 1, 2, 3, 4\}$. Let us now consider all possible cases.

For $m = 0$, equation (1) has two integer solutions, namely $x_1 = 0$ and $x_2 = 8$.

This yields two solutions of the given problem: $(x, y) \in \{(0, 0), (8, 8)\}$.

For $m \in \{-3, -2, -1, 1, 2, 3\}$, equation (1) has no integer solutions.

For $m = 4$, equation (1) has two integer solutions, namely $x_3 = 0$ and $x_4 = 4$. By virtue of $y = x + m$, the given problem has two more integer solutions: $(x, y) \in \{(0, 4), (4, 8)\}$. Because the given equation is symmetric, we know that if a pair (x, y) is a solution, then (y, x) is also solution. We therefore obtain further solutions $(x, y) \in \{(4, 0), (8, 4)\}$, and this completes the final case.

The given equation has six integer solutions, namely

$$(x, y) \in \{(0, 0), (8, 8), (0, 4), (4, 8), (4, 0), (8, 4)\}.$$

Problem 51. (A-I-4-99) *Determine all integer solutions of the equation*

$$(x^2 - y^2)^2 = 16y + 1.$$

Solution: Let (x, y) be an integer solution of the given equation. The left-hand side of the equation is non-negative, and we therefore have $16y+1 \geq 0$.

Furthermore, because y is an integer, this implies $y \geq 0$.

Let us now consider two possible cases:

Case (a): $x^2 - y^2 \geq 0$. Because $y \geq 0$ holds, and x is only represented in the equation in an even power, we may assume with loss of generality that x is non-negative, and therefore there exists a number $m \geq 0$, such that $x = y + m$.

For $m = 0$, we obtain $y = x$, and the pair (x, x) is not a solution of the given equation.

Substituting $x = y + m$ with $m \geq 1$ in the given equation, we obtain

$$(y^2 + 2my + m^2 - y^2)^2 = 16y + 1,$$

whence

$$m^2(2y + m)^2 = 16y + 1.$$

Because $m^2 \geq 1$ holds, this implies

$$(2y + m)^2 \leq 16y + 1 \iff 4y^2 + 4(m - 4)y + m^2 - 1 \leq 0.$$

The discriminant of the quadratic function on the left side of this inequality is equal to $\Delta = 16(17 - 8m)$, and we know that a solution of the inequality can only exist if $\Delta \geq 0$, i.e., $17 - 8m \geq 0$, holds. Because m is an integer with $m \geq 1$, it follows that $m = 1$ or $m = 2$. This gives us two possibilities to consider:

Case (a1): For $m = 1$ we have $x = y + 1$ and after substituting in the equation, we have

$$((y+1)^2 - y^2)^2 = 16y + 1, \quad \text{or} \quad 4y^2 - 12y = 0,$$

yielding $y_1 = 0$ and $y_2 = 3$.

For $y_1 = 0$, we obtain $x_1 = 1$ and we have the first solution $(1, 0)$. If we have the solution (x, y), then $(-x, y)$ is also a solution, giving us $(-1, 0)$ as another solution of the given equation.

For $y_2 = 3$, we have $x_2 = 4$, which gives us two more solutions, namely $(4, 3)$ and $(-4, 3)$.

Case (a2): For $m = 2$ we have $x = y + 2$ and after substituting in the equation, we have

$$((y+2)^2 - y^2)^2 = 16y + 1, \quad \text{or} \quad 16y^2 + 16y + 16 = 1.$$

This is a contradiction, because the left-hand side of the last equation is divisible by 16 and right-hand side is certainly not.

Case (b): $x^2 - y^2 < 0$. As the argument used in Case (a) also applies here, we again obtain the two possible values $y_1 = 0$ and $y_2 = 3$.

Case (b1): For $y_1 = 0$ the equation gives us $(x^2 - y^2)^2 = 1$, and because $x^2 - y^2$ is negative, we have $x^2 - y^2 = -1$ and therefore $x^2 = -1$, which is not possible.

Case (b2): For $y_2 = 3$ the equation gives us $(x^2 - y^2)^2 = 49$, and because $x^2 - y^2$ is negative, we have $x^2 - y^2 = -7$ and therefore $x^2 = 2$. This is a contradiction to the assumption that x is an integer. We see that this case does not yield any further solutions.

In summary, we see that the equation has the four solutions

$$(x, y) \in \{(-1, 0), (1, 0), (4, 3), (-4, 3)\}. \qquad \square$$

Problem 52. (A-T-1-98) *Determine all pairs (x, y) of integers fulfilling the equation*

$$x^3 - 4x^2 - 5x = 6^y.$$

Solution: Let us consider three possible cases:

(a) If $y < 0$, the left-hand side is an integer and the right-hand side is not. The given equation has no solutions in this case.

(b) If $y = 0$, we obtain the algebraic equation $x^3 - 4x^2 - 5x - 1 = 0$. If an integer solution of this equation exists, it must be a divisor of -1. Neither 1 nor -1 are roots of the polynomial, as can easily be confirmed. Once again, the given equation does not have any integer solutions in this case.

(c) If $y > 0$, we write the left-hand side of the equation in the form

$$(x - 5)x(x + 1).$$

Because the right-hand side is a positive number, we have $x(x - 5)(x + 1) > 0$. The values of x fulfilling this inequality are $x \in (-1, 0) \cup (5, \infty)$. Because x must be an integer, this means $x \in [6, \infty)$.

We now consider two subcases:

Case (c1): The number x is even. Then the numbers $x - 5$ and $x + 1$ are both odd. These must therefore both be powers of 3, and the difference of these powers of 3 is equal to 6. The successive powers of the number 3 are: $1, 3, 9, 27, 81, \ldots$ and the difference of two successive numbers from the above sequence is an increasing sequence. Because the difference is equal to 6, this is possible only if we have $3^2 = 9$ and $3^1 = 3$. Then $x - 5 = 3$ and $x + 1 = 9$ and we have $x = 8$. This gives us the product $(x - 5)x(x + 1) = 216 = 6^3$, and we obtain the solution $x = 8$ and $y = 3$.

Case (c2): The number x is odd. Then the numbers $x - 5$ and $x + 1$ are both even. These must therefore both be powers of 2, and the difference of these powers of 2 is equal to 6. The successive powers of the number 2 are: $1, 2, 4, 8, 16, 32, \ldots$ and the difference of two successive numbers from the above sequence is an increasing sequence. Because the difference is equal to 6, this is possible only if we have $2^3 = 8$ and $2^1 = 2$. Then $x - 5 = 2$ and $x + 1 = 8$ and we have $x = 7$. This gives us the product $(x - 5)x(x + 1) = 112$, which is not an integer power of 6. There are therefore no integer solutions in this case.

In conclusion. we see that the unique solution of the given equation is $x = 8$, $y = 3$. □

Problem 53. (B-T-1-97) *Determine all triples* (x, y, z) *of positive integers fulfilling the equation*

$$\frac{1}{x} + \frac{2}{y} + \frac{3}{z} = 4.$$

Solution: Let us consider four possible cases:

(a) If $z = 1$, the given equation can be written as $\frac{1}{x} + \frac{2}{y} = 1$. This certainly implies $y > 2$. For $y = 3$ we obtain $x = 3$ and for $y = 4$ we obtain $x = 2$. For $y \geq 5$ we have $\frac{1}{x} = 1 - \frac{2}{y}$, or $\frac{1}{x} = \frac{y-2}{y}$, which implies

$$x = \frac{y}{y-2} = 1 + \frac{2}{y-2}.$$

This is certainly never an integer, and we see that there are no solutions for $y \geq 5$.

Case (a) therefore yields the two positive integer solutions (2, 4, 1) and (3, 3, 1).

(b) If $z = 2$, the given equation can be written as $\frac{1}{x} + \frac{2}{y} = \frac{5}{2}$. For $y = 1$ we obtain $x = 2$. For $y \geq 2$ we have $\frac{1}{x} = \frac{5}{2} - \frac{2}{y} = \frac{2y}{5y-4}$, which implies

$$x = \frac{2y}{5y-4} < 1.$$

This is again certainly never an integer, and we see that there are no solutions for $y \geq 2$.

Case (b) therefore yields one further positive integer solution, namely (2, 1, 2).

(c) If $z = 3$, the given equation can be written as $\frac{1}{x} + \frac{2}{y} = 3$. For $y = 1$ we obtain $x = 1$. For $y \geq 2$ we have $\frac{2}{y} \leq 1$ and $\frac{1}{x} \leq 1$ and the sum $\frac{1}{x} + \frac{2}{y}$ can certainly not be equal to 3.

Case (c) therefore yields one more positive integer solution, namely (1, 1, 3).

(d) If $z \geq 4$, we have $\frac{3}{z} \leq \frac{3}{4}$, $\frac{2}{y} \leq 2$, and $\frac{1}{x} \leq 1$, and the sum $\frac{1}{x} + \frac{2}{y} + \frac{3}{z}$ is certainly not equal to 4. Case (d) therefore yields no positive integer solutions.

The given equation therefore has four positive integer solutions, namely

$$(x, y, z) \in \{(2, 4, 1), (3, 3, 1), (2, 1, 2), (1, 1, 3)\}. \quad \square$$

Problem 54. (B-T-1-96) *Determine all integer solutions of the equation*

$$xyz + xy + yz + xz + x + y + z = 1996.$$

Solution: After adding 1 to both sides of the equation, factorising gives us

$$(x + 1)(y + 1)(z + 1) = 1997.$$

Because 1997 is a prime number, the factorisation of 1997 is only possible in the form

$$1997 = 1997 \cdot 1 \cdot 1 = 1997 \cdot (-1) \cdot (-1) = (-1997) \cdot (-1) \cdot 1.$$

These three factors are $x+1$, $y+1$ and $z+1$. From this, we immediately obtain the set of 12 integer solutions (x, y, z):

$$\begin{aligned}&\{(1996, 0, 0), (0, 1996, 0), (0, 0, 1996), (1996, -2, -2),\\ &\quad(-2, 1996, -2), (-2, -2, 1996), (1998, -2, 0), (-2, 1998, 0),\\ &\quad(-2, 0, 1998), (1998, 0, -2), (0, 1998, -2), (0, -2, 1998)\}.\end{aligned}$$ □

Problem 55. (B-T-3-95) *Determine as many triples of different positive integers (k, l, m) as possible, fulfilling the equation*

$$\frac{1}{1995} = \frac{1}{k} + \frac{1}{l} + \frac{1}{m}.$$

Solution: The Diophantine equation

$$\frac{1}{x} + \frac{1}{y} + \frac{1}{z} = 1$$

has the solutions (2, 3, 6), (2, 4, 4), (3, 3, 3) in positive integers, along with their permutations, yielding a total of 10 solutions. Writing the equation in the form

$$1 = \frac{1995}{x \cdot 1995} + \frac{1995}{y \cdot 1995} + \frac{1995}{z \cdot 1995},$$

and dividing both sides of the equation by 1995 therefore gives us 10 solutions of the given equation.

In order to find further solutions, we note that $1995 = 3 \cdot 5 \cdot 7 \cdot 19$. The number 1995 therefore has 16 positive integer divisors: 1, 3, 5, 7, 15, 19, 21, 35, 57, 95, 105, 133, 285, 399, 665 and 1995. Let a, b be two divisors of the number 1995 and $n = 1995 + a + b$. We can show that the numbers $k = n, l = n \cdot \frac{1995}{a}, m = \frac{1995}{b}$ are then also positive integer solutions of the given equation.

Substitution in the given equation yields

$$\frac{1}{1995} = \frac{1}{n} + \frac{1}{n \cdot \frac{1995}{a}} + \frac{1}{n \cdot \frac{1995}{b}}$$

or

$$\frac{1}{1995} = \frac{1}{n} + \frac{a}{n \cdot 1995} + \frac{b}{n \cdot 1995}.$$

Multiplying both sides of this equation by $n \cdot 1995$, we obtain the established identity $n = 1995 + a + b$, and this is certainly true. □

Problem 56. (A-T-2-94) *Determine all pairs* (x, y) *of integers fulfilling the equation*

$$\sqrt{x} + \sqrt{y} = \sqrt{90}.$$

Solution: We note that $x, y \geq 0$ must hold. Furthermore, we also know that $x, y \leq 90$ must hold.

From the given equation we have $\sqrt{x} = \sqrt{90} - \sqrt{y}$ and after squaring both sides of this equation, we obtain $x = 90 - 2\sqrt{90y} + y$, or $\sqrt{360y} = 90 + y - x$. On the right-hand side of this equation, we have a non-negative integer. We therefore have $360y = k^2$, where k is a non-negative integer. Since the left side of this expression is divisible by 36, the right side must be as well. Therefore, we have $6|k$ and $k = 6 \cdot m$, where m is a non-negative integer. After substitution in the last equation, we have $360y = 36 \cdot m^2$, and we therefore obtain $10y = m^2$. The left-hand side is divisible by 10, which implies $10|m^2$, and it follows that $10|m$ must hold. We can therefore write $m = 10n$, where n is a non-negative integer and $10y = 100n^2$, or $y = 10n^2$. Because $0 \leq y \leq 90$, n can only take values from the set $\{0, 1, 2, 3\}$ and y from the set $\{0, 10, 20, 30\}$. Calculating the corresponding values of x from $x = 90 - 2\sqrt{90y} + y$, we obtain all pairs of solutions, namely $\{(0, 90), (10, 40), (40, 10), (90, 0)\}$. □

2.5. Systems of Diophantine Equations

An interesting variation on the subject results from the consideration of more than one Diophantine equation. Such problems combine the flavour of Diophantine equations with that of systems of equations, resulting in an interesting amalgam of number theoretical and algebraic structures.

PROBLEMS

57. Solve the following system of equations in the domain of integers:

$$x + \frac{2}{y} = z,$$

$$y + \frac{4}{z} = x,$$

$$z - \frac{6}{x} = y.$$

58. Determine all positive integers n for which there exist positive integers x and y satisfying

$$x + y = n^2,$$

$$10x + y = n^3.$$

59. Determine all integer solution of the following system of equations:

$$x^2z + y^2z + 4xy = 40,$$

$$x^2 + y^2 + xyz = 20.$$

SOLUTIONS

Problem 57. (A-I-1-12) *Solve the following system of equations in the domain of integers*:

$$x + \frac{2}{y} = z,$$

$$y + \frac{4}{z} = x,$$

$$z - \frac{6}{x} = y.$$

Solution: Note that the fraction $\frac{2}{y}$ is an integer and therefore $y \in \{-2, -1, 1, 2, \}$. Since $x, y, z \neq 0$, we can multiply the equations by y, z and x, respectively, obtaining the system

$$xy + 2 = yz,$$
$$yz + 4 = zx,$$
$$zx - 6 = xy.$$

It is easy to see that the last equation results from the first two and can therefore be omitted.

We now observe that if a triple of integers (x, y, z) is a solution of the system, the triple $(-x, -y, -z)$ is then also a solution. Thus, we have to consider only two cases for $y > 0$.

If $y = 1$, we get

$$x + 2 = z,$$
$$z + 4 = zx.$$

Substituting for z in the second equation and solving the resulting quadratic equation gives us two solutions: $z = -1$, $x = -3$ and $z = 4$, $x = 2$.

If $y = 2$, we get

$$2x + 2 = 2z,$$
$$2z + 4 = zx.$$

Once again, substituting for z in the second equation and solving the resulting quadratic equation gives us two solutions: $z = -1$, $x = -2$ and $z = 4$, $x = 3$.

The complete solution therefore consists of eight triples (x, y, z):

$$(-3, 1, -1), (2, 1, 4), (-2, 2, -1), (3, 2, 4), (3, -1, 1),$$
$$(-2, -1, -4), (2, -2, 1), (-3, -2, -4)$$ □

Problem 58. (B-I-2-08) *Determine all positive integers n for which there exist positive integers x and y satisfying*

$$x + y = n^2,$$
$$10x + y = n^3.$$

Solution: From the first equation, we have $y = n^2 - x$, and substituting in the second equation, we obtain $9x = n^2(n - 1)$. The left side of this is divisible by 3.

Let us now consider two possible cases:
Case (a): n is divisible by 3, i.e., $n = 3k$ for $k \in \mathbb{Z}^+$. Then we have

$$x = k^2(3k - 1) = 3k^3 - k^2 \quad \text{and} \quad y = (3k^2 - (3k^3 - k^2) = 10k^2 - 3k^3.$$

Because $y \geq 1$, we have $10k^2 - 3k^3 \geq 1$ and $k^2(10 - 3k) \geq 1$, and it follows that $10 - 3k \geq 1$, i.e., $k \leq 3$.
Then $k \in \{1, 2, 3\}$ and $n \in \{3, 6, 9\}$. For these values of n, the corresponding x and y are positive integers.

Case (b): $n - 1$ is divisible by 3. Then n is not divisible by 3 and $n - 1$ is divisible by 9. We obtain $n - 1 = 9m$ for some non-negative integer m, and therefore $n = 9m + 1$. We then have

$$9x = (9m - 1)^2 \cdot 9m, \quad \text{or} \quad x = m(9m + 1)^2,$$

and

$$y = n^2 - x, \quad \text{or} \quad y = (9m + 1)^2(1 - m).$$

Because $y \geq 1$, we have $(9m + 1)^2(1 - m) \geq 1$, and it follows that $m \leq 0$. This is impossible, because x is a positive integer.

Another solution: As above, we have $x = \frac{1}{9}n^2(n - 1)$. We can consider three possible cases:

Case (a): $n = 3k$, $k \in \mathbb{Z}$. Then $x = k^2(3k - 1) \in \mathbb{Z}$.
Case (b): $n = 3k + 1$, $k \in \mathbb{Z}$. Then $x = 3k^3 + k^2 - \frac{3k+1}{9} \notin \mathbb{Z}$.
Case (c): $n = 3k + 2$, $k \in \mathbb{Z}$. Then $x = 3k^3 + 3k^2 - \frac{4}{9} \notin \mathbb{Z}$.

We must therefore have $n = 3k$, where k is a positive integer. From the first solution, we have $x = k^2(3k - 1)$ and therefore $y = k^2(10k - 3)$. Because $y \geq 1$, we have $1 - 3k \geq 1$ and therefore $k \geq 3$. It follows that $k \in \{1, 2, 3\}$ and $n \in \{3, 6, 9\}$ must hold.

Another solution: From the first solution, we have $x = \frac{1}{9}n^2(n - 1)$ and x is a positive integer. It then follows that $n - 1 \geq 1$, and therefore $n \geq 2$, must hold. Therefore, we have $y = \frac{n^2}{9}(10 - n)$. From $y \geq 1$ it then follows that $10 - n \geq 1$, i.e., $n \leq 9$, holds.

It can easily be verified that for an integer n, $2 \leq n \leq 9$, only $n \in \{3, 6, 9\}$ yield positive integers for x and y, and we see that the system only has solutions in positive integers for $n \in \{3, 6, 9\}$. □

Problem 59. (A-I-3-05) *Determine all integer solution of the following system of equations*:

$$x^2z + y^2z + 4xy = 40,$$
$$x^2 + y^2 + xyz = 20.$$

Solution: Multiplying both sides of the second equation by 2, we obtain the system

$$x^2z + y^2z + 4xy = 40,$$
$$2x^2 + 2y^2 + 2xyz = 40,$$

and adding these equations gives us

$$x^2z + y^2z + 4xy + 2x^2 + 2y^2 + 2xyz = 80,$$

or

$$(x + y)^2(z + 2) = 80. \tag{1}$$

Because $80 = 5 \cdot 2^4$, there are now three possible cases:

(a) $x + y = \pm 1$. Then by (1), we have $(\pm 1)^2(z + 2) = 80$ and $z + 2 = 80$, or $z = 78$. Substituting $z = 78$ in the given system, we obtain

$$78(x^2 + y^2) + 4xy = 40,$$
$$x^2 + y^2 + 78xy = 20,$$

or

$$39(x^2 + y^2) + 2xy = 20,$$
$$x^2 + y^2 = 20 - 78xy.$$

Substituting $x^2 + y^2$ from the second equation in the first, we obtain

$$39(20 - 78xy) + 2xy = 20 \iff 304xy = 76,$$

which is not possible for integer values of x, y.

(b) $x + y = \pm 2$. Then by (1), we have $(\pm 2)^2(z + 2) = 80$ and $z + 2 = 20$ or $z = 18$. Substituting $z = 18$ in the given system, we obtain

$$18(x^2 + y^2) + 4xy = 40,$$
$$x^2 + y^2 + 18xy = 20,$$

or

$$1(x^2 + y^2) + 2xy = 20,$$
$$x^2 + y^2 = 20 - 18xy.$$

Substituting $x^2 + y^2$ from the second equation in the first, we obtain

$$9(20 - 18xy) + 2xy = 20 \iff xy = 1.$$

Since x and y are integers, this implies $x_1 = y_1 = 1$ or $x_2 = y_2 = -1$. Checking the triples $(x, y, z) \in \{(1, 1, 18), (-1, -1, 18)\}$, we see that both are solutions of the given system.

(c) $x + y = \pm 4$. Then by (1), we have $(\pm 4)^2(z + 2) = 80$ and $z + 2 = 5$ or $z = 3$. Substituting $z = 3$ in the given system, we obtain

$$3(x^2 + y^2) + 4xy = 40,$$
$$x^2 + y^2 + 3xy = 20,$$

or

$$3(x^2 + y^2) + 4xy = 40,$$
$$x^2 + y^2 = 20 - 3xy.$$

Substituting $x^2 + y^2$ from the second equation in the first, we obtain

$$3(20 - 3xy) + 4xy = 40 \iff xy = 4,$$

and the second equation then gives us

$$x^2 + y^2 = 20 - 3xy = 20 - 3 \cdot 4 = 8.$$

This means that we have

$$x^2 - 2xy + y^2 = 8 - 2 \cdot 4 = 0, \quad \text{or} \quad (x - y)^2 = 0,$$

and therefore $x = y$.

Because either $x + y = 4$ or $x + y = -4$ holds, this implies either $x_3 = y_3 = 2$ or $x_4 = y_4 = -2$. Checking the triples $(x, y, z) \in \{(2, 2, 3), (-2, -2, 3)\}$, we see that these are also solutions of the given system.

The given system of Diophantine equations therefore has four solutions, namely $(x, y, z) \in \{(1, 1, 18), (-1, -1, 18), (2, 2, 3), (-2, -2, 3)\}$. □

2.6. Systems of Diophantine Inequalities

A subject another step beyond systems of Diophantine equations is that of systems of such inequalities. While systems of regular inequalities generally have infinite numbers of solutions, this need not be the case for Diophantine inequalities. This means that finding the solution to such a problem can involve elements not just from solving systems of equations and number theoretical ideas, but also some basic ideas on inequalities.

PROBLEMS

60. Determine how many triples (x, y, z) of positive integers satisfy the following system of inequalities

$$\frac{x^2}{x^2 + 2yz} + \frac{y^2}{y^2 + 2zx} + \frac{z^2}{z^2 + 2xy} \leq 1,$$
$$x^2 + y^2 + z^2 \leq 2001.$$

61. Determine all pairs (x, y) of integers fulfilling the following system of inequalities:

$$x^2 + x + y \leq 3,$$
$$y^2 + y + x \leq 3.$$

SOLUTIONS

Problem 60. (B-I-4-01) *Determine how many triples* (x, y, z) *of positive integers satisfy the following system of inequalities:*

$$\frac{x^2}{x^2+2yz}+\frac{y^2}{y^2+2zx}+\frac{z^2}{z^2+2xy}\leq 1,$$
$$x^2+y^2+z^2\leq 2001.$$

Solution: Using the well-known inequalities

$$x^2+y^2\geq 2xy,\quad x^2+z^2\geq 2xz,\quad y^2+z^2\geq 2yz,$$

which are true for any positive integers x, y, z we obtain

$$\frac{x^2}{x^2+2yz}+\frac{y^2}{y^2+2zx}+\frac{z^2}{z^2+2xy}$$
$$\geq\frac{x^2}{x^2+y^2+z^2}+\frac{y^2}{x^2+y^2+z^2}+\frac{z^2}{x^2+y^2+z^2}=1.$$

From the first inequality of the given system of inequalities, it therefore follows that

$$\frac{x^2}{x^2+2yz}+\frac{y^2}{y^2+2zx}+\frac{z^2}{z^2+2xy}=1$$

must hold for any solutions of the given system. Because we also have

$$\frac{x^2}{x^2+y^2+z^2}+\frac{y^2}{x^2+y^2+z^2}+\frac{z^2}{x^2+y^2+z^2}=1,$$

subtracting these last two equalities gives us

$$x^2\left(\frac{1}{x^2+2yz}-\frac{1}{x^2+y^2+z^2}\right)+y^2\left(\frac{1}{y^2+2xz}-\frac{1}{x^2+y^2+z^2}\right)$$
$$+z^2\left(\frac{1}{z^2+2xy}-\frac{1}{x^2+y^2+z^2}\right)=0,$$

or

$$x^2\frac{y^2+z^2-2yz}{(x^2+2yz)(x^2+y^2+z^2)}+y^2\frac{x^2+z^2-2xz}{(y^2+2xz)(x^2+y^2+z^2)}$$
$$+z^2\frac{x^2+y^2-2xy}{(y^2+2xy)(x^2+y^2+z^2)}=0.$$

The sum of these three non-negative fractions is equal to 0, and therefore each of these fractions must be equal to 0. We therefore have

$$x^2+y^2-2xy=0,\quad x^2+z^2-2xz=0,\quad\text{and}\quad y^2+z^2-2yz=0,$$

which implies $x=y=z$.

From the second inequality of the given system, we therefore obtain the condition $3x^2 \leq 2\,001$, i.e., $x^2 \leq 667$. This inequality is fulfilled for any integer x from the set $\{0, 1, \ldots, 25\}$.

The number of positive integer solutions of the given system of inequalities is therefore 25. Specifically, these are all triples in the form (x, x, x), where $x \in \{1, 2, \ldots, 25\}$. □

Problem 61. (B-I-2-94) *Determine all pairs* (x, y) *of integers fulfilling the following system of inequalities*:

$$x^2 + x + y \leq 3,$$
$$y^2 + y + x \leq 3.$$

Solution: From the first inequality, we obtain $y \leq -x^2 - x + 3$. Pairs of integers (x, y) fulfilling this inequality can be interpreted as lattice points lying not above the parabola with the equation $y = -x^2 - x - 3$.

Similarly, the second inequality can be written as $x \leq -y^2 - y + 3$. Pairs of integers (x, y) fulfilling this inequality can be interpreted as lattice points lying not to the right of the parabola with the equation $x = -y^2 - y - 3$. The situation described here is illustrated in the figure.

In order to find the points of intersection of the two parabolas, we must solve the system of equations given by

$$x^2 + x + y = 3$$
$$y^2 + y + x = 3.$$

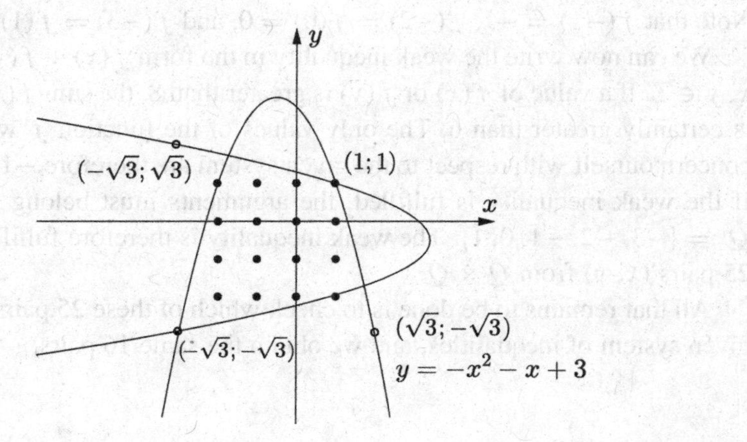

Subtracting these equations, we obtain $x^2 - y^2 = 0$ and therefore $y = x$ or $y = -x$. Substituting $y = x$ in the first equation gives us the equation $x^2 + 2x - 3 = 0$ with the roots $x_1 = 1$ and $x_2 = -3$. We have therefore found two points of intersection, namely $(1, 1)$ and $(-3, -3)$.

Similarly, substituting $y = -x$ in the first equation gives us $x^2 - 3 = 0$, and therefore $x_3 = \sqrt{3}$ and $x_4 = -\sqrt{3}$. This gives us the other two points of intersection, namely: $(\sqrt{3}, \sqrt{3})$ and $(-\sqrt{3}, -\sqrt{3})$.

Both inequalities are fulfilled by the lattice points lying in the common part of the areas fulfilling each inequality.

The lattice points in this common area are the following:

$$(-3,-3), (-2,0), (-2,-1), (-2,-2), (-1,1), (-1,0),$$
$$(-1,-1), (-1,-2), (0,1), (0,0), (0,-1), (0,-2),$$
$$(1,1), (1,0), (1,-1), (1,-2).$$

These are precisely the required integer solutions of the system of inequalities.

Another solution: Adding both sides of the inequalities, we obtain the weaker inequality $x(x+2) + y(y+2) \leq 6$. This inequality is weaker because more pairs fulfilling this inequality exists than those fulfilling the given system of inequalities.

We can now write $f(a) = a^2 + 2a = (a+1)^2 - 1$. From this, we see that the inequality $f(a) \geq 1$ holds for all values of a, and for integer values of a, the function takes on integer values from the set $\{-1, 0, 3, 8, 15, \dots\}$. Note that $f(-1) = -1$, $f(-2) = f(0) = 0$, and $f(-3) = f(1) = 2$.

We can now write the weak inequality in the form $f(x) + f(y) \leq 6$ for $x, y \in \mathbb{Z}$. If a value of $f(x)$ or $f(y)$ is greater than 8, the sum $f(x) + f(y)$ is certainly greater than 6. The only values of the function f we need to concern ourself with respect to the given system are therefore -1, 0, and 3. If the weak inequality is fulfilled, the arguments must belong to the set $Q := \{-3, -2, -1, 0, 1\}$. The weak inequality is therefore fulfilled by the 25 pairs (x, y) from $Q \times Q$.

All that remains to be done is to check which of these 25 pairs fulfil the given system of inequalities, and we obtain the same 16 pairs. $\square$

Chapter 3

Algebra

3.1. Algebraic Equations

When students start to participate in mathematical competitions, many of them expect to encounter problems similar to those they are used to in standard school-book style, just a bit harder. In this situation, solving equations is certainly an expected type of activity. Of course, each equation they encounter at competition level introduces some new twist, and their problem solving creativity is put to the test again and again.

PROBLEMS

62. Determine all pairs (a, b) of real numbers such that the roots of the cubic equation

$$x^3 + ax^2 + bx + ab = 0$$

are the numbers $-a, -b$ and $-ab$.

63. Let $a \neq 0, b, c$ be real numbers with $|a + c| < |a - c|$. Prove that the quadratic equation $ax^2 + bx + c = 0$ has two real roots, one of which is positive and one of which is negative.

64. Let a be an arbitrary real number. Prove that real numbers b and c certainly exist, such that

$$\sqrt{a^2 + b^2 + c^2} = a + b + c$$

65. Determine all cubic polynomials $P(x)$ with real coefficients such that the equation $P(x) = 0$ has three real roots (not necessarily different) fulfilling the following conditions:

(a) The number 1 is a root of the considered equation.
(b) For each root t of the equation $P(x) = 0$ the condition $P(2t) = t$ holds.

66. Determine all triples of mutually distinct real numbers a, b, c such that the cubic equation

$$x^3 + abx^2 + bcx + ca = 0$$

with unknown x has three real roots a, b, c.

67. We are given two real numbers x and y $(x \neq y)$ such that $x^4 + 5x^3 = y$ and $y^3 + 5x^2 = 1$ hold. Prove that the equality $x^3 + x^2y + xy^2 = -1$ holds.

68. Suppose the quadratic equation

$$x^2 - 2007x + b = 0$$

with a real parameter b has two positive integer roots. Determine the maximum value of b.

69. Determine all real solutions (x, y) of the equation

$$2 \cdot 3^{2x} - 2 \cdot 3^{x+1} + 3^x \cdot 2^{y+1} + 2^{2y} - 2^{y+2} + 5 = 0.$$

SOLUTIONS

Problem 62. (A-I-1-15) *Determine all pairs (a, b) of real numbers such that the roots of the cubic equation*

$$x^3 + ax^2 + bx + ab = 0$$

are the numbers $-a$, $-b$ and $-ab$.

Solution: Let us suppose that there exist real numbers a, b fulfilling the conditions of the given problem. Using Vièta's formulas we get

$$-a^2b^2 = -ab \Leftrightarrow ab(ab - 1) = 0, \tag{1}$$

$$ab + a^2b + ab^2 = b \Leftrightarrow ab(1 + a + b) = b, \tag{2}$$

$$-a - b - ab = -a \Leftrightarrow b(a + 1) = 0. \tag{3}$$

We now consider two different cases for factors on the left side of (1):

First, let $ab = 0$. If $a = 0$, then by (2) and (3) $b = 0$ also holds. If $b = 0$, then similarly by (2) and (3) a can be an arbitrary real number. This means that desired pairs (a, b) of real numbers are $(a, 0)$, where a is any real number. The corresponding cubic equation is in the form $x^3 + ax^2 = 0$, which has three real roots $-a$, 0 and 0.

Now let $ab = 1$. Since $b \neq 0$, then $a = -1$ must hold by (3). From the initial condition $ab = 1$ we further have $b = -1$. It is easy to check that the pair $(a, b) = (-1; -1)$ of real numbers also satisfies the relation (2). In this case, we obtain the cubic equation $x^3 - x^2 - x + 1 = 0$, which has three real roots -1, -1 and 1.

In summary, we see that the solutions of the given problem are the following pairs of real numbers: $(a, b) = (a, 0)$, where a is any real number, and also $(a, b) = (-1; -1)$. □

Problem 63. (A-I-1-14) *Let $a \neq 0$, b, c be real numbers with $|a + c| < |a - c|$. Prove that the quadratic equation $ax^2 + bx + c = 0$ has two real roots, one of which is positive and one of which is negative.*

Solution: Since both sides of the inequality $|a + c| < |a - c|$ are non-negative real numbers with $a \neq 0$, we equivalently get (after squaring of both sides and easy manipulation) the inequality $ac < 0$ and also $c/a < 0$. This implies that also inequality $b^2 - 4ac > 0$ holds, i.e., the discriminant of the quadratic equation $ax^2 + bx + c = 0$ is a positive real number. Therefore this equation has two real roots x_1 and x_2.

Moreover, applying Vièta's formula we obtain $x_1x_2 = c/a < 0$, which means that the real roots x_1 and x_2 have opposite signs and the proof is finished. □

Problem 64. (A-I-1-13) *Let a be an arbitrary real number. Prove that real numbers b and c certainly exist such that*

$$\sqrt{a^2 + b^2 + c^2} = a + b + c$$

Solution: The equality

$$\sqrt{a^2 + b^2 + c^2} = a + b + c$$

is equivalent to the conditions

$$a^2 + b^2 + c^2 = (a + b + c)^2 \quad \text{and} \quad a + b + c \geq 0, \tag{1}$$

and the first of these is equivalent to

$$ab + bc + ca = 0.$$

Now, let a be an arbitrary real number. Let us choose $b \neq 0$ and c such that

$$a + b > 0 \quad \text{and} \quad c = -\frac{ab}{a+b}.$$

The last equality yields $ab + bc + ca = 0$ and

$$a + b + c = a + b - \frac{ab}{a+b} = \frac{a^2 + ab + b^2}{a+b}.$$

Since the discriminant D of the trinomial $x^2 + bx + b^2$ is equal to $-3b^2 < 0$, so for $x = a$ we get $a^2 + ab + b^2 > 0$.

Thus we have shown that, for an arbitrary a, there exist b and c that fulfil the conditions (1). □

Note: Another way to solve this problem is to consider the following. For arbitrary $a \geq 0$ one can take $b = c = 0$ and for arbitrary $a < 0$ the given equality is fulfilled for $b = c = -2a$.

Problem 65. (A-I-3-12) *Determine all cubic polynomials $P(x)$ with real coefficients such that the equation $P(x) = 0$ has three real roots (not necessarily different) fulfilling the following conditions*:

(a) *The number* 1 *is a root of the considered equation.*
(b) *For each root t of the equation $P(x) = 0$ the condition $P(2t) = t$ holds.*

Solution: If 1 is a triple root of $P(x) = 0$ then $P(x) = a(x-1)^3$ and from the condition (b) we obtain $a \cdot 1^3 = 1$ and therefore $a = 1$. In this case we therefore have

$$P(x) = (x-1)^3. \tag{1}$$

If 1 is a double root of $P(x) = 0$ then $P(x) = a(x-1)^2(x - x_1)$ where $x_1 \neq 1$. Using (b) for $t = 1$ and $t = x_1$ we obtain

$$a(2 - x_1) = 1 \quad \text{and} \quad a(2x_1 - 1)^2 x_1 = x_1.$$

From the first equation $a \neq 0$ and $x_1 = 2 - \frac{1}{a}$. The second equation gives us either $x_1 = 0$ (so $a = \frac{1}{2}$) or $(2x_1 - 1)^2 = \frac{1}{a}$. Substituting for $x_1 = 2 - \frac{1}{a}$

we get

$$9 - \frac{13}{a} + \frac{4}{a^2} = 0.$$

Solving this equation we obtain $a = 1$ and $a = \frac{4}{9}$.

In these cases $x_1 = 1$ (which contradicts $x_1 \neq 1$) or $x_1 = -\frac{1}{4}$. We therefore obtain further solutions

$$P(x) = \tfrac{1}{2}x(x-1)^2 \quad \text{and} \quad P(x) = \tfrac{4}{9}(x-1)^2(x+\tfrac{1}{4}). \tag{2}$$

If 1 is a single root of $P(x) = 0$ and this equation has only one root $x_1 \neq 1$ then $P(x) = a(x-1)(x-x_1)^2$. Using (b) we obtain

$$a(2-x_1)^2 = 1 \quad \text{and} \quad a(2x_1 - 1)x_1^2 = x_1.$$

From the first equation, we have $a \neq 0$ and $\frac{1}{a} = (2-x_1)^2$. The second equation implies either $x_1 = 0$ (so $a = \frac{1}{4}$) or $2x_1^2 - x_1 = \frac{1}{a}$. Substituting for $\frac{1}{a} = (2-x_1)^2$ we get

$$x_1^2 + 3x_1 - 4 = 0.$$

Solving this equation, we obtain $x_1 = 1$ (which contradicts $x_1 \neq 1$) and $x_1 = -4$, so $a = \frac{1}{36}$. In this case we get further solutions

$$P(x) = \tfrac{1}{4}x^2(x-1) \quad \text{and} \quad P(x) = \tfrac{1}{36}(x-1)(x+4)^2. \tag{3}$$

Finally, let us assume that the cubic equation has three distinct roots and $P(x) = ax^3 + bx^2 + cx + d$.

From (b) it follows that an equation $P(2x) - x = 0$ has the same three distinct roots. This implies that the equation

$$0 = 8P(x) - (P(2x) - x) = 4bx^2 + (6c+1)x + 7d$$

has the same three distinct roots and so its coefficients vanish. We easily get $b = 0$, $c = -\frac{1}{6}$, $d = 0$. The condition (a) implies that 1 is a root and so $a - \frac{1}{6} = 0$. From this, it follows that $a = \frac{1}{6}$ and in this case we get the solution

$$P(x) = \tfrac{1}{6}(x^3 - x) = \tfrac{1}{6}(x-1)\,x\,(x+1). \tag{4}$$

The given equation thus has exactly the six solutions given by (1)–(4). □

Problem 66. (A-I-1-10) *Determine all triples of mutually distinct real numbers a, b, c such that the cubic equation*

$$x^3 + abx^2 + bcx + ca = 0$$

with unknown x has three real roots a, b, c.

Solution: Using Vieta's formulas we must determine the real numbers a, b, c such that the equations

$$a + b + c = -ab, \quad ab + bc + ca = bc \quad \text{and} \quad abc = -ca$$

hold. After easy rewriting of these equations we have

$$a(b + 1) + (b + c) = 0, \quad a(b + c) = 0 \quad \text{and} \quad ac(b + 1) = 0.$$

If $a = 0$, then $b + c = 0$ and we obtain the solution $(a, b, c) = (0, b, -b)$ with an arbitrary real b.

If $a \neq 0$, then both $b + c = 0$ and $b + 1 = 0$ must be fulfilled. This implies $b = -1$ and $c = 1$. Thus, we obtain the further solution $(a, b, c) = (a, -1, 1)$ with arbitrary real $a \neq 0$.

In conclusion, we see that the set of all solutions of the given problem consists of the triples in the form $(0, -b, b)$ with $b \neq 0$ together with all triples in the form $(a, -1, 1)$ with $a \neq -1, a \neq 0$ and $a \neq 1$. □

Problem 67. (A-T-1-10) *We are given two real numbers x and y $(x \neq y)$ such that $x^4 + 5x^3 = y$ and $y^3 + 5x^2 = 1$ hold. Prove that the equality $x^3 + x^2y + xy^2 = -1$ holds.*

Solution: Multiplying $y^3 + 5x^2 = 1$ by x yields $xy^3 + 5x^3 = x$. Taking the difference of this and the first equation $x^4 + 5x^3 = y$ we obtain

$$x(x^3 - y^3) = y - x,$$

and dividing this equation by $x - y \neq 0$ yields

$$x(x^2 + xy + y^2) = -1,$$

which is equivalent to

$$x^3 + x^2y + xy^2 = -1,$$

as claimed. □

Problem 68. (A-I-1-07) *Suppose the quadratic equation*

$$x^2 - 2007x + b = 0$$

with a real parameter b has two positive integer roots. Determine the maximum value of b.

Solution: If the given quadratic equation has two roots, the discriminant D must fulfil the property $D = 2007^2 - 4b > 0$, and hence $b < 2007^2/4$ must hold. Then there exist two different roots x_1, x_2 of the equation.

By Vieta's formulas we have $x_1 + x_2 = 2007$ and $x_1 \cdot x_2 = b$. By the first formula we have $x_2 = 2007 - x_1$.

Then $b = x_1 \cdot x_2 = x_1 \cdot (2007 - x_1) = -x_1^2 + 2007x_1$ holds for the coefficient b.

Let us consider the function $f(x) = -x^2 + 2007x$ for $x \in \mathbb{R}$. It is a quadratic function which takes its maximum value for $x_0 = \frac{2007}{2}$, but x_0 is not an integer.

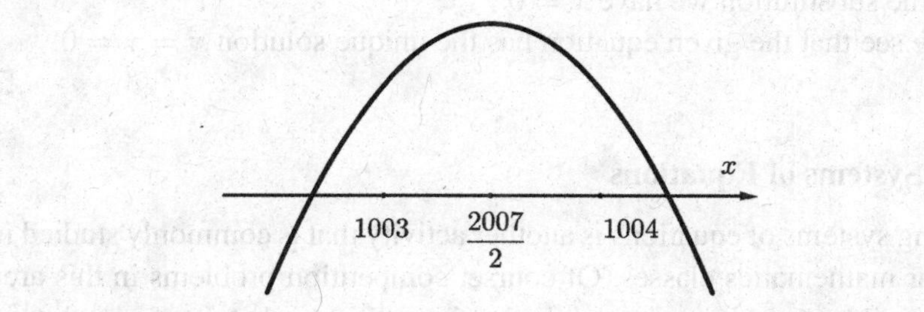

As we see in the above figure, the function f assumes its maximal value in the nearest positive integer points, i.e., $x_1 = 1003$ or $x_2 = 1004$. We note that

$$f(1003) = f(1004) = 1\,007\,012.$$

Then we have two numbers such that the function f takes its maximum: 1003 and 1004. By $x_2 = 2007 - x_1$ we obtain that the second number is equal to 1004 or 1003, respectively. The maximal value is therefore $b = x_1 \cdot x_2 = 1003 \cdot 1004 = 1\,007\,012$. □

Problem 69. (A-T-1-07) *Determine all real solutions* (x, y) *of the equation*

$$2 \cdot 3^{2x} - 2 \cdot 3^{x+1} + 3^x \cdot 2^{y+1} + 2^{2y} - 2^{y+2} + 5 = 0.$$

Solution: Rewriting the equation in an equivalent form, we obtain

$$2 \cdot (3^x)^2 - 6 \cdot 3^x + 2 \cdot 3^x \cdot 2^y + (2^y)^2 - 4 \cdot 2^y + 5 = 0.$$

Substituting $u = 3^x$, $v = 2^y$ in the above equation, we obtain

$$2u^2 + (2v - 6)u + v^2 - 4v + 5 = 0.$$

The last expression is a quadratic equation with unknown u and with the parameter v. The discriminant of this equation is

$$D = -4(v - 1)^2.$$

If the equation has a real solution, the discriminant must have the property $D \geq 0$, which is equivalent to $-4(v - 1)^2 \geq 0$. Hence, it is possible only if $D = 0$ and this equality holds only for $v = 1$. From the substitution $2^y = 1$, and we see that $y = 0$.

If $D = 0$, the quadratic equation has the unique double root $u = 1$, and from the substitution we have $x = 0$.

We see that the given equation has the unique solution $x = y = 0$.

□

3.2. Systems of Equations

Solving systems of equations is another activity that is commonly studied in regular mathematics classes. Of course, competition problems in this area usually add some kind of an original twist to the usual style of questions.

PROBLEMS

70. We are given the following system of equations:

$$x + y + z = a,$$
$$x^2 + y^2 + z^2 = b^2,$$

with real parameters a and b. Prove that the system of equations has a solution in real numbers if and only if the inequality

$$|a| \leq |b|\sqrt{3}$$

holds.

71. Solve the following system of equations in the domain of real numbers:

$$x^4 + 1 = 2yz,$$
$$y^4 + 1 = 2xz,$$
$$z^4 + 1 = 2xy.$$

72. Solve the following system of equations in real numbers:

$$\sqrt{\sqrt{x} + 2} = y - 2,$$
$$\sqrt{\sqrt{y} + 2} = x - 2.$$

73. Determine all triples (x, y, z) of real numbers satisfying the following system of equations:

$$x^2 + y = z^3 + 1,$$
$$y^2 + z = x^3 + 1,$$
$$z^2 + x = y^3 + 1.$$

74. Determine all positive real numbers x, y and z for which the system of equations

$$x(x + y) + z(x - y) = 65,$$
$$y(y + z) + x(y - z) = 296,$$
$$z(z + x) + y(z - x) = 104.$$

is fulfilled.

75. Let $n \geq 3$ be a positive integer. Solve the following (cyclic) system of equations in non-negative real numbers:

$$2x_1 + x_2^2 = x_3^3$$
$$2x_2 + x_3^2 = x_4^3$$
$$\vdots$$
$$2x_{n-1} + x_n^2 = x_1^3$$
$$2x_n + x_1^2 = x_2^3.$$

76. Solve the following system of equations in real numbers:

$$x + 3yz = 24,$$
$$y + 3zx = 24,$$
$$z + 3xy = 24.$$

77. Solve the following system of equations in real numbers:

$$\sqrt{x^2 + (y+1)^2} + \sqrt{(x-12)^2 + (4-y)^2} = 13,$$
$$5x^2 - 12xy = 24.$$

78. Determine all possible positive values of the product pqr if real numbers p, q, r are given for which

$$pq + q + 1 = qr + r + 1 = rp + p + 1 = 0$$

holds.

79. Solve the following system of equations in real numbers with non-negative real parameters p, q, r satisfying $p + q + r = \frac{3}{2}$:

$$x = p^2 + y^2,$$
$$y = q^2 + z^2,$$
$$z = r^2 + x^2.$$

SOLUTIONS

Problem 70. (A-T-1-13) *We are given the following system of equations*:

$$x + y + z = a,$$
$$x^2 + y^2 + z^2 = b^2.$$

with real parameters a and b. Prove that the system of equations has a solution in real numbers if and only if the inequality

$$|a| \leq |b|\sqrt{3}$$

holds.

Solution: We first assume that a solution (x, y, z) of the given system of equations in real numbers exists. Squaring the first equation, we obtain the equivalent system

$$x^2 + y^2 + z^2 + 2xy + 2xz + 2yz = a^2,$$
$$x^2 + y^2 + z^2 = b^2.$$

Applying the inequality $2x^2 + 2y^2 + 2z^2 \geq 2xy + 2xz + 2yz$, for which equality holds only for $x = y = z$, we obtain the system

$$3x^2 + 3y^2 + 3z^2 \geq a^2,$$
$$x^2 + y^2 + z^2 = b^2,$$

from which $3b^2 \geq a^2$ follows, which is equivalent to $\sqrt{3}|b| \geq |a|$.

In order to prove the second part of the claim, we assume that the inequality $3b^2 \geq a^2$ holds. We will prove that a solution (x, y, z) of the given system of equations in real numbers must exist in this case.

We shall prove that there exists such a solution of the given system with $z = x$. In order to do this, we consider the resulting system of equations

$$2x + y = a,$$
$$2x^2 + y^2 = b^2.$$

Substituting $y = a - 2x$ from the first equation in the second equation, we obtain the quadratic equation

$$6x^2 - 4ax + a^2 - b^2 = 0$$

with the real parameters a, b. The discriminant of this equation is

$$D = 8(3b^2 - a^2),$$

and from the assumption $3b^2 \geq a^2$ there certainly exists a real root x of the equation. Therefore, real values for $y = a - 2x$ and $z = x$ also exist in this case, and the proof is complete. □

Geometric solution. This is an especially instructive problem, since it allows multiple solutions of very different types. Having considered a purely algebraic solution, we now consider a geometric interpretation.

The first equation $x + y + z = a$ can be interpreted as the equation of the plane containing the points $A(a, 0, 0)$, $B(0, a, 0)$ and $C(0, 0, a)$, as in the figure below. (For $a = 0$ this is a plane passing through the origin.) If O is the origin, $ABCO$ is a tetrahedron (degenerate for $a = 0$) with edge-lengths

$$|AB| = |BC| = |CA| = a\sqrt{2} \text{ and } |AO| = |BO| = |CO| = a.$$

Similarly, the second equation $x^2 + y^2 + z^2 = b^2$ can be interpreted as the equation of a sphere with the centre in O and radius $|b|$.

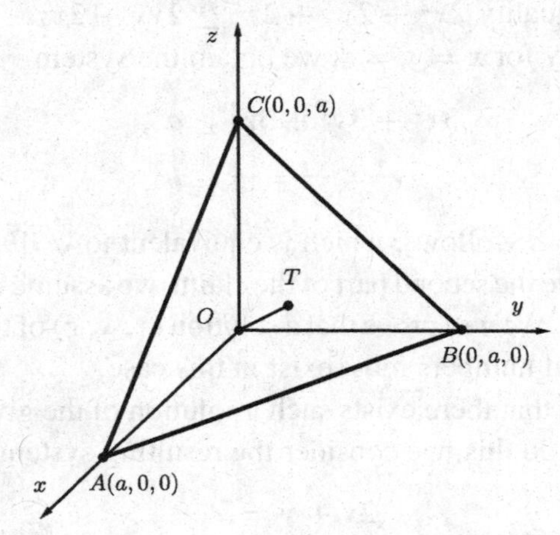

Let T be the centroid of the face ABC of the tetrahedron $ABCO$ and let $|OT| = d$. (Note that $a = 0$ implies $d = 0$.) It is easy to see that the segment OT is the altitude of this tetrahedron from the vertex O. We can now compute the volume V of the tetrahedron $ABCO$ in two ways, obtaining

$$V = \tfrac{1}{6}\,|a|^3 = \tfrac{1}{3}\,P \cdot d,$$

where P is the area of the face ABC. P is easily calculated as

$$P = \tfrac{1}{2}\,|a|\sqrt{2} \cdot |a|\sqrt{\tfrac{3}{2}} = \tfrac{1}{2}\,a^2\sqrt{3}$$

and we therefore have

$$V = \tfrac{1}{6}\,|a|^3 = \tfrac{1}{6}\,a^2\sqrt{3} \cdot d,$$

which implies

$$d = \tfrac{\sqrt{3}}{3}\,|a|.$$

Finally, the given system of equations with unknowns x, y, z (and real parameters a, b) has a real solution if and only if the inequality $d \le |b|$

holds, i.e.,

$$\frac{\sqrt{3}}{3}|a| \leq |b|.$$

This inequality is equivalent to $|a| \leq |b|\sqrt{3}$ which concludes the proof. □

Another solution: Tools from inequality theory give us yet another method to attack this problem.

Let a triple (x_1, x_2, x_3) of real numbers be a solution of the given system of equations. Using the Cauchy–Schwarz inequality, we get

$$3b^2 = (1^2 + 1^2 + 1^2)(x_1^2 + x_2^2 + x_3^2) \geq (x_1 + x_2 + x_3)^2 = a^2, \qquad (1)$$

i.e., $a^2 \leq 3b^2$ and therefore $|a| \leq |b|\sqrt{3}$.

Conversely, we can assume that $|a| \leq |b|\sqrt{3}$ holds. This implies that there exist real numbers x_1, x_2, x_3 solving the given system of equations because of inequality (1), and the proof is finished. □

Problem 71. (A-I-1-11) *Solve the following system of equations in the domain of real numbers*:

$$x^4 + 1 = 2yz,$$

$$y^4 + 1 = 2xz,$$

$$z^4 + 1 = 2xy.$$

Solution: For any real a, the inequality $2a^2 \leq a^4 + 1$ certainly holds. We therefore obtain the following estimates for the left-hand sides of all of the equations of the given system:

$$2x^2 \leq x^4 + 1 = 2yz,$$

$$2y^2 \leq y^4 + 1 = 2zx,$$

$$2z^2 \leq z^4 + 1 = 2xy.$$

Adding all three inequalities (and dividing by 2) we have

$$x^2 + y^2 + z^2 \leq xy + yz + zx. \qquad (1)$$

On the other hand we know that the inequality

$$x^2 + y^2 + z^2 \geq xy + yz + zx \qquad (2)$$

is true with equality if and only if $x = y = z$ holds. It therefore only remains to solve the biquadratic equation $x^4 + 1 = 2x^2$. It is easy to see that this equation has only two real roots, namely 1 and -1.

After checking, we therefore see that the given system of equations has only two real solutions, namely $(x, y, z) = (1, 1, 1)$ and $(x, y, z) = (-1, -1, -1)$. □

Remark. There is an interesting alternative way to prove $x = y = z$. Multiplying the given equations by x, y and z respectively, we obtain

$$x^5 + x = 2xyz,$$

$$y^5 + y = 2xyz,$$

$$z^5 + z = 2xyz.$$

Therefore we have $x^5+x = y^5+y = z^5+z$. Since the function $f(t) = t^5+t$ is increasing in the whole domain (as the sum of two increasing functions), equality holds if and only if $x = y = z$.

Problem 72. (A-I-4-10) *Solve the following system of equations in real numbers*:

$$\sqrt{\sqrt{x} + 2} = y - 2,$$

$$\sqrt{\sqrt{y} + 2} = x - 2.$$

Solution: Since $\sqrt{\sqrt{x} + 2} \geq \sqrt{2}$ certainly holds, the first equation gives us $y \geq 2 + \sqrt{2}$ and the second similarly gives us $x \geq 2 + \sqrt{2}$. After squaring both given equations, we obtain

$$\sqrt{x} + 2 = (y - 2)^2, \tag{1}$$

$$\sqrt{y} + 2 = (x - 2)^2. \tag{2}$$

After subtraction, we obtain

$$\sqrt{x} - \sqrt{y} = (y - x)(x + y - 4),$$

which is equivalent to

$$(\sqrt{x} - \sqrt{y})\big(1 + (\sqrt{x} + \sqrt{y})(x + y - 4)\big) = 0.$$

The second factor on the left side of this equation is always greater than zero due to the estimates found at the beginning of the solution. It therefore follows that $x = y = t^2$ with $t \geq \sqrt{2 + \sqrt{2}}$, and thus $t > 1$.

From (1) and (2) it now follows that t must satisfy the condition

$$t+2=(t^2-2)^2 \iff t^4-4t^2-t+2=(t-2)(t^3+2t^2-1)=0.$$

Since $t^3+2t^2-1>1+2-1=2$ holds for each $t>1$, the last algebraic equation (of the fourth degree) has only one root $t>1$, namely $t=2$ and we therefore have $x=y=4$. The given system of equations thus has only one solution, namely $(x,y)=(4,4)$, in real numbers. □

Remark. The proof of the equality $x=y$ $(x\geq 2, y\geq 2)$ can also be shown in the following way: If $x>y$, then

$$\sqrt{x}+2=(y-2)^2<(x-2)^2=\sqrt{y}+2, \quad \text{i.e., } x<y,$$

which contradicts $x>y$. Since the analogous contradiction follows if we assume $x<y$, $x=y$ must hold.

Problem 73. (A-I-1-09) *Determine all triples* (x,y,z) *of real numbers satisfying the following system of equations*:

$$x^2+y=z^3+1,$$
$$y^2+z=x^3+1,$$
$$z^2+x=y^3+1.$$

Solution: First, we shall consider the case when $x=y=z$. In this case, we obtain the equation

$$x^2+x=x^3+1 \iff (x+1)(x-1)^2=0.$$

The roots of this equation are $x_1=-1$ and $x_2=1$. In this manner, we obtain two *constant* solutions of the system: $(x,y,z)=(1,1,1)$ and $(x,y,z)=(-1,-1,-1)$.

Next, we shall prove that no other (not constant) solutions (x,y,z) of the given system of equations exist.

In order to do this, we define the function

$$f(u)=u^3-u^2-u+1 \quad \text{for } u\in\mathbb{R}.$$

The graph of the function f is shown in the following figure:

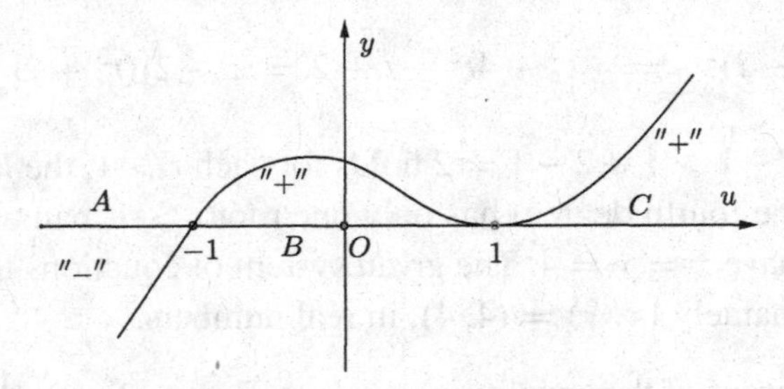

We now define the following intervals on the real line:

$$A = (-\infty, -1], \quad B = (-1, 1] \quad \text{and} \quad C = (1, +\infty).$$

The inequality $f(u) \geq 0$ has the solutions $u \in B \cup C$ and the inequality $f(u) \leq 0$ has the solutions $u \in A$.

We can now approach the complete solution of the problem step by step by considering some properties of the solution of the given system in the following lemmas.

Lemma 1. *Let* (x, y, z) *be a solution of the given system. If* $x, y, z \in \{-1, 1\}$, *then* $x = y = z = -1$ *or* $x = y = z = 1$.

Proof. From the assumptions, we know that at least two of the numbers x, y, z are certainly equal. If $x = y = 1$, the second equation $1^2 + z = 1^3 + 1$ gives us $z = 1$, and if $x = y = -1$, it similarly gives us $(-1)^2 + z = (-1)^3 + 1 \iff z = -1$. Analogous results are obtained for $x = z$ and $y = z$. □

Lemma 2. *The condition*

$$f(x) + f(y) + f(z) = 0$$

is necessary for the existence of a solution of the given system.

Proof. Adding the three equations of the given system gives us this equality, since we then obtain

$$x + y + z + x^2 + y^2 + z^2 = x^3 + y^3 + z^3 + 3$$
$$\iff f(x) + f(y) + f(z) = 0.$$
□

Lemma 3. *If* $x, y, z \in A$ *and* (x, y, z) *are not all equal to* -1, *then* (x, y, z) *is not a solution of the given system.*

Proof. If $x, y, z \in A$, the inequalities

$$f(x) \le 0, \quad f(y) \le 0, \quad f(z) \le 0$$

must hold. Because (x, y, z) are not all equal to -1, at least one of these inequalities is strong by virtue of Lemma 1. We then have

$$f(x) + f(y) + f(z) < 0,$$

and by Lemma 2 the triple (x, y, z) is not a solution of the given system. □

Lemma 4. *If $x, y, z \in B \cup C$ and (x, y, z) are not all equal to 1, then (x, y, z) is not a solution of the given system.*

Proof. If $x, y, z \in B \cup C$, the inequalities

$$f(x) \ge 0, \quad f(y) \ge 0, \quad f(z) \ge 0$$

must hold. Because (x, y, z) are not all equal to 1, at least one of these inequalities is strong by virtue of Lemma 1. We then have

$$f(x) + f(y) + f(z) > 0,$$

and by Lemma 2 the triple (x, y, z) is not a solution of the given system. □

Lemma 5. *If exactly two of the numbers x, y, z belong to the set A, then (x, y, z) is not a solution of the given system.*

Proof. If $x, y \in A$, we have $x^3 \le -1$ and $-y^2 \le -1$. Adding these inequalities, we obtain $x^3 - y^2 \le -2$, or $x^3 - y^2 + 1 \le -1$. By the second equation, we therefore obtain $z = x^3 - y^2 + 1 \le -1$, and therefore $z \in A$. By Lemma 3, the triple (x, y, z) is therefore not a solution of the given system. The analogous argument holds for $x, z \in A$ or $y, z \in A$. □

Lemma 6. *If exactly two of the numbers x, y, z belong to the set B, then (x, y, z) is not a solution of the given system.*

Proof. If $x, y \in B$, then $x^3 \in [-1, 1]$ and $-y^2 \in [-1, 0]$. Adding these inequalities, we obtain $x^3 - y^2 \in [-2, 1]$ and $x^3 - y^2 + 1 \in [-1, 2]$. By the second equation, we therefore obtain $z \in [-1, 2]$, and therefore $z \in B \cup C$. We therefore have a situation in which $x, y, z \in B \cup C$ with x, y, z not all equal to 1 (since $x, y \in B$ but $z \notin B$), and by Lemma 4, the triple (x, y, z) is therefore not a solution of the given system. The analogous argument holds for $x, z \in B$ or $y, z \in B$. □

Lemma 7. *If exactly two of the numbers x, y, z belong to the set C, then (x, y, z) is not a solution of the given system.*

Proof. If $x, y \in C$, then $x^2 \geq 1$ and $y \geq 1$. Adding these inequalities, we obtain $x^2 + y \geq 2$ and therefore $x^2 + y - 1 \geq 1$. By the first equation, we therefore obtain $z^3 = x^2 + y - 1 \geq 1$, or $z \in C$. We therefore have $x, y, z \in B \cup C$ with x, y, z not all equal to 1 (since $x, y \in C$ but $z \notin C$), and by Lemma 4, the triple (x, y, z) is therefore not a solution of the given system. The analogous argument holds for $x, z \in C$ or $y, z \in C$. □

Lemma 8. *If every number x, y, z belongs to exactly one of the intervals A, B, C, then (x, y, z) is not a solution of the given system.*

Proof. We consider all possible cases:

Case I: If $x \in A$ and $y \in C$, then $x^2 \geq 1$ and $y \geq 1$. Adding these inequalities, we obtain $x^2 + y \geq 2$ and therefore $x^2 + y - 1 \geq 1$. By the first equation, we have $z^3 \geq 1$, i.e., $z \in C$ or $z = 1$. Since we have assumed $z \in B$, we must have $z = 1$. The first equation then gives us $x^2 + y = 2$. Because $x \in A \Rightarrow x^2 \geq 1$ and $y \in C \Rightarrow y \geq 1$, we therefore have $x^2 = 1$ and $y = 1$. Because $y = z = 1$, we therefore have $x = y = z = 1$ by Lemma 1, contradicting the assumption that x, y, z lie in the three separate intervals A, B, C.

Case II: If $x \in C$ and $y \in A$, then $-x \leq -1$ and $y^2 \leq -1$. Adding these inequalities, we obtain $-x + y^3 \leq -2$ and $-x + y^3 + 1 \leq -1$. By the third equation, we therefore have $z^2 \leq -1$ and the number z cannot exist.

Contradictions also result from all remaining cases: Case III: $x \in A$ and $z \in C$, Case IV: $x \in C$ and $z \in A$, Case V: $y \in A$ and $z \in C$, and Case VI: $z \in A$ and $y \in C$ by analogous arguments. □

We are now ready to summarise. In Lemmas 1–8, we have considered all possible cases for the numbers x, y, z in the intervals A, B, C. The system of equations therefore has exactly two solutions, namely $(x, y, z) = (-1, -1, -1)$ and $(x, y, z) = (1, 1, 1)$. □

Problem 74. (A-I-2-07) *Determine all positive real numbers x, y and z for which the system of equations*

$$x(x+y)+z(x-y)=65,$$
$$y(y+z)+x(y-z)=296,$$
$$z(z+x)+y(z-x)=104$$

is fulfilled.

Solution: Adding the first and third equations, we obtain $z^2+2zx+x^2=169$, and therefore $(z+x)^2=13^2$.

Adding the first two equations, we obtain $x^2+2xy+y^2=361$, and therefore $(x+y)^2=19^2$.

Adding the last two equations, we obtain $y^2+2yz+z^2=400$, and therefore $(y+z)^2=20^2$.

Because x, y, z are positive numbers, we therefore have $x+y=19$, $y+z=20$ and $z+x=13$. Adding these three equations and dividing by 2, we obtain $x+y+z=26$.

Successively subtracting the equations $x+y=19$, $y+z=20$ and $z+x=13$ from this equation therefore gives us $x=6$, $y=13$ and $z=7$. Checking shows us that this is indeed a solution of the given system.

The given system of equations therefore has the unique solution $(x, y, z)=(6, 13, 7)$. □

Problem 75. (A-T-3-07) *Let $n \geq 3$ be a positive integer. Solve the following (cyclic) system of equations in non-negative real numbers*:

$$2x_1+x_2^2=x_3^3$$
$$2x_2+x_3^2=x_4^3$$
$$\vdots$$
$$2x_{n-1}+x_n^2=x_1^3$$
$$2x_n+x_1^2=x_2^3.$$

Solution: First, we consider the case when $x_1=x_2=\cdots=x_n=u$. We obtain the equation

$$2u+u^2=u^3 \iff u(u^2-u-2)=0.$$

This equation has three roots: $u_1=0$, $u_2=2$, $u_3=-1$. Because the solution $(x_1, x_2, \ldots, x_n)$ is non-negative, this yields two solutions:

$x_1 = x_2 = \cdots = x_n = 0$ and $x_1 = x_2 = \cdots = x_n = 2$ of the given system of equations.

We will now prove that no other solutions of the given system of equations exist.

For this purpose, we define the two intervals $A = [0, 2]$ and $B = [2, \infty)$. Considering the function $f(u) = u(u^2 - u - 2)$, we note that the inequality $u(u^2 - u - 2) \leq 0$ holds for $u \in A$ and the inequality $u(u^2 - u - 2) \geq 0$ holds for $u \in B$, as we can see in the following figure.

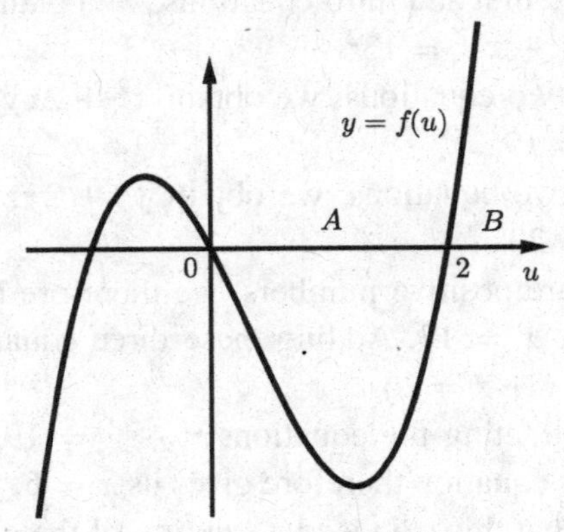

Adding all of the equations of the given system, we can write the sum in the form

$$(x_1^3 - x_1^2 - 2x_1) + (x_2^3 - x_2^2 - 2x_2) + \cdots + (x_n^3 - x_n^2 - 2x_n) = 0.$$

In the case when all $x_1, x_2, \ldots, x_n \in A$, and at least one of the numbers x_i, $i = 1, 2, \ldots, n$, belong to the open interval $(0, 2)$, we have $x_i^3 - x_i^2 - 2x_i < 0$ and the above expression is not equal to 0. In this case, we must therefore have $x_1, x_2, \ldots, x_n \in \{0, 2\}$.

It is not possible for a solution to exist, in which the values of the x_i are not all simultaneously either 0 or 2. If this were the case, one of the equations would be true if we insert one of the triples

$$(0, 0, 2), (0, 2, 0), (2, 0, 0), (2, 0, 2), (0, 2, 2) \quad \text{or} \quad (2, 2, 0),$$

and it is easy to check that this cannot be the case, yielding the required contradiction. We therefore see that the only solutions for which all values of $(x_1, x_2, \ldots, x_n)$ lie in A are those for which all values of the x_i are either 0 or 2.

Next, we consider the case in which all numbers $x_1, x_2, \dots x_n$ lie in B. If at least one of the numbers x_i, $i = 1, 2, \dots, n$ is greater than 2, we have $x_i^3 - x_i^2 - 2x_i > 0$. Once again, the above expression is not equal to 0. In this case, we therefore certainly have $x_1 = x_2 = \dots = x_n = 2$.

Now, we shall prove that any solution $(x_1, x_2, \dots, x_n)$ must have the following property: If two successive terms $x_i, x_{i+1} \in A$ for some $i \in \{1, \dots, n\}$ (modulo n), then the next term x_{i+2} must also lie in A. For this purpose, let us consider the equation in which x_i, x_{i+1} occur on the left-hand side of the equation. Without loss of generality, we assume that this is the first equation. We assume $x_1, x_2 \in A$. Then we have $2x_1 \le 4$ and $x_2^2 \le 4$. Therefore we also have $2x_1 + x_2^2 = x_3^3 \le 8$, and therefore $x_3 \le 2$ and $x_3 \in A$.

Similarly, if we have a solution $(x_1, x_2, \dots, x_n)$ in which two successive terms $x_i, x_{i+1} \in B$, the next term x_{i+2} must also lie in B. Let us consider the equation in which x_i, x_{i+1} occur on the left hand side of the equation. Again, without loss of generality, we assume that this is the first equation. We then have $x_1, x_2 \in B$. This gives us $2x_1 \ge 4$ and $x_2^2 \ge 4$ and therefore $2x_1 + x_2^2 = x_3^3 \ge 8$, which yields $x_3 \ge 2$ and therefore $x_3 \in B$.

We see that the only possible solutions in which two successive terms x_i, x_{i+1} belong to the same set A or B are those for which all terms are either equal to 0 or 2. Any other solutions that may exist must therefore alternate the values of successive terms between the intervals A and B.

For odd values of n, the cyclic nature of this property means that two successive terms must lie in the same interval, and this is therefore not possible.

The only case that remains is the case for even n in which the values of successive terms alternate between the intervals A and B. In this case, let us consider three successive terms x_i, x_{i+1}, x_{i+2} (cyclic) and a triple with values $x_i, x_{i+2} \in B$ and $x_{i+1} \in A$. Due to the cyclic nature of the assumed structure, we can assume without loss of generality that $x_i \le x_{i+2}$ holds. As before, we can also assume that the considered triple is from the first equation of the system. We then have $x_1, x_3 \in B$, $x_2 \in A$ and $x_1 \le x_3$. Let $x_3 = x_1 + d$, where $d \ge 0$. Then, from the first equation, we have

$$2x_1 + x_2^2 = (x_1 + d)^3 \ge x_1^3, \quad \text{and therefore} \quad x_2^2 \ge x_1(x_1^2 - 2) \ge 4.$$

for $x_1 \ge 2$. This gives us $x_2 \ge 2$ and therefore $x_2 \in B$, in contradiction to the assumption. This last case is therefore also not possible.

We see that the system has exactly two non-negative solutions: $x_1 = x_2 = \cdots = x_n = 0$ and $x_1 = x_2 = \cdots = x_n = 2$. □

Problem 76. (A-T-1-06) *Solve the following system of equations in real numbers*:

$$\begin{aligned} x + 3yz &= 24, \\ y + 3zx &= 24, \\ z + 3xy &= 24. \end{aligned}$$

Solution: The system of equations is symmetric, and the set of solutions is therefore also symmetric.

Subtracting the second equation from the first, we obtain

$$x - y + 3yz - 3zx = 0, \quad \text{which implies } (x - y)(1 - 3z) = 0.$$

Let us now consider two possible cases:

Case I: $y = x$. Substituting in the second and third equation we have

$$\begin{aligned} x + 3zx &= 24, \\ z + 3x^2 &= 24. \end{aligned}$$

Subtracting these equations, we obtain

$$x - z + 3xz - 3x^2 = 0 \iff (x - z)(1 - 3x) = 0.$$

We therefore have

$$z = x \quad \text{or} \quad x = \tfrac{1}{3}.$$

If $z = x$, the second equation gives us $3x^2 + x - 24 = 0$ with the roots $x_1 = -3$ and $x_2 = \frac{8}{3}$. We therefore obtain the two solutions $x_1 = y_1 = z_1 = -3$ and $x_2 = y_2 = z_2 = \frac{8}{3}$. These solutions indeed fulfil the given system.

If $x = \frac{1}{3}$, substitution gives us $z = 23\frac{2}{3}$. With this, we have a third solution of the given system, namely $x_3 = y_3 = \frac{1}{3}$, $z_3 = 23\frac{2}{3}$. This solution also fulfils the given system.

Case II: $1 - 3z = 0$, or $z = \frac{1}{3}$. Substituting $z = \frac{1}{3}$ in the second and third equation we obtain the system of equations

$$\begin{aligned} x + y &= 24, \\ \tfrac{1}{3} + 3xy &= 24. \end{aligned}$$

Substituting $y = 24 - x$ from the first equation in the second, we obtain the quadratic equation $3x^2 - 72x + \frac{71}{3} = 0$ with the roots $x_4 = \frac{1}{3}$ and $x_5 = 23\frac{2}{3}$. This gives us the last two solutions $x_4 = \frac{1}{3}$, $y_4 = 23\frac{2}{3}$, $z_4 = \frac{1}{3}$ and $x_5 = 23\frac{2}{3}$, $y_5 = \frac{1}{3}$, $z_5 = \frac{1}{3}$ of the given system of equations.

The given system of equations has exactly five solutions (x, y, z):

$$(-3, -3, -3), \quad (2\tfrac{2}{3}, 2\tfrac{2}{3}, 2\tfrac{2}{3}), \quad (\tfrac{1}{3}, \tfrac{1}{3}, 23\tfrac{2}{3}), \quad (\tfrac{1}{3}, 23\tfrac{2}{3}, \tfrac{1}{3}), \quad (23\tfrac{2}{3}, \tfrac{1}{3}, \tfrac{1}{3}).$$

□

Problem 77. (A-I-1-03) *Solve the following system of equations in real numbers:*

$$\sqrt{x^2 + (y+1)^2} + \sqrt{(x-12)^2 + (4-y)^2} = 13,$$
$$5x^2 - 12xy = 24.$$

Solution: We consider the following points in the plane: $A(0, -1)$, $B(12, 4)$ and $P(x, y)$. Then $|AB| = 13$,

$$|AP| = \sqrt{x^2 + (y+1)^2}, \quad \text{and} \quad |BP| = \sqrt{(x-12)^2 + (y-4)^2}.$$

By the first equation we get $|AP| + |PB| = |AB|$ and the point P therefore lies on the segment AB.

The equation of the line passing through the two points A and B is $y = \frac{5}{12}x - 1$. Because the point P lies on the segment AB, we have $0 \leq x \leq 12$. Substituting $y = \frac{5}{12}x - 1$ in the second equation, we have

$$5x^2 - 12x(\tfrac{5}{12}x - 1) = 24, \quad \text{or } 5x^2 - 5x^2 + 12x = 24.$$

This yields $x = 2$, and substituting $x = 2$ into the first equation we have $y = -\frac{1}{6}$.

The given system therefore has the unique solution $(x, y) = (2, -\frac{1}{6})$.

□

Problem 78. (A-I-2-06) *Determine all possible positive values of the product pqr if real numbers p, q, r are given for which*

$$pq + q + 1 = qr + r + 1 = rp + p + 1 = 0$$

holds.

Solution: Multiplying the three given equations by r, p, q respectively, we obtain

$$pqr = -r(q+1),$$

$$pqr = -p(r+1),$$
$$pqr = -q(p+1).$$

Multiplying all of these equations gives us

$$(pqr)^3 = -pqr(p+1)(q+1)(r+1).$$

The original system of equations also gives us

$$-1 = (-1)^3 = (pq+q)(qr+r)(rp+p) = pqr(p+1)(q+1)(r+1),$$

and comparing the right-hand sides of the last two equations, we obtain the equation $(pqr)^3 = 1$, and therefore $pqr = 1$.

Note that there does indeed exist a solution of the system, namely $p = -2, q = 1, r = -\frac{1}{2}$. □

Problem 79. (A-I-2-16) *Solve the following system of equations in real numbers with non-negative real parameters* p, q, r *satisfying* $p+q+r = \frac{3}{2}$:

$$x = p^2 + y^2,$$
$$y = q^2 + z^2,$$
$$z = r^2 + x^2.$$

Solution: For any non-negative parameters p, q, r we have (by the QM–AM inequality)

$$\sqrt{\frac{p^2+q^2+r^2}{3}} \geq \frac{p+q+r}{3} \Leftrightarrow p^2+q^2+r^2 \geq \frac{(p+q+r)^2}{3} \geq \frac{1}{3}\cdot\frac{9}{4} = \frac{3}{4}$$

with equality if and only if $p = q = r = \frac{1}{2}$. Adding all three equations, we obtain

$$\begin{aligned} 0 &= x^2+y^2+z^2-(x+y+z)+(p^2+q^2+r^2) \\ &\geq x^2+y^2+z^2-(x+y+z)+\frac{3}{4} \\ &= \left(x-\frac{1}{2}\right)^2 + \left(y-\frac{1}{2}\right)^2 + \left(z-\frac{1}{2}\right)^2 \geq 0 \end{aligned}$$

for $p = q = r = \frac{1}{2}$.

After checking, we see that the triple $(x, y, z) = \left(\frac{1}{2}, \frac{1}{2}, \frac{1}{2}\right)$ is the solution of the given system of equations with $p+q+r = \frac{3}{2}$. □

Another solution: Adding all three equations, we obtain

$$x + y + z = p^2 + q^2 + r^2 + x^2 + y^2 + z^2,$$

which is equivalent to

$$-p^2 - q^2 - r^2 = x(x - 1) + y(y - 1) + z(z - 1).$$

We will prove that the minimum of the right-hand side of this equation is equal to $-\frac{3}{4}$ and the maximum of the left-hand side is equal to $-\frac{3}{4}$. Equality then holds if and only if the right-hand side and left-hand side are both equal to $-\frac{3}{4}$.

Let us consider the function $f(t) = t(t - 1)$ for $t > 0$. The minimum of f is equal to $-\frac{1}{4}$. It follows that the minimum of the right-hand side is equal to $-\frac{3}{4}$, and this holds for $x = y = z = \frac{1}{2}$.

Because $p + q + r = \frac{3}{2}$, we have $p^2 + q^2 + r^2 + 2pq + 2pr + 2qr = \frac{9}{4}$. By the well-known inequality $p^2 + q^2 + r^2 \geq pq + pr + qr$ we therefore obtain

$$p^2 + q^2 + r^2 \geq \frac{3}{4}, \text{ i.e., } -p^2 - q^2 - r^2 \leq -\frac{3}{4}$$

and the maximum of the left-hand side is equal to $-3/4$ if and only if $p = q = r = \frac{1}{2}$.

After checking, we see that the triple $(x, y, z) = (\frac{1}{2}, \frac{1}{2}, \frac{1}{2})$ is the unique solution of the given system of equations.

Remark. If we replace the assumption $p + q + r = \frac{3}{2}$ by the inequality $p + q + r \geq \frac{3}{2}$, the new generalised problem has the same solution as the above problem.

3.3. Functional Equations

A type of problem not usually considered in standard mathematics classes concerns the solution of functional equations. In some interesting ways, these problems can be considered to be precursors to the important topic of differential equations in applied mathematics, but high school students will generally only be confronted with this type of problem if they are somehow involved in the world of mathematics competitions.

PROBLEMS

80. Let $f : \mathbb{R} \to \mathbb{R}$ be a real function satisfying

$$f(f(x) - y) = x - f(y) \quad \text{for all } x, y \in \mathbb{R}.$$

Prove that f is an odd function, i.e., $f(-x) = -f(x)$ holds for all $x \in \mathbb{R}$.

81. Let $\mathbb{R}^+ = (0; +\infty)$. Determine all functions $f : \mathbb{R}^+ \to \mathbb{R}$, such that

$$xf(x) = xf\left(\frac{x}{y}\right) + yf(y)$$

holds for all positive real values of x and y.

82. Determine all functions $f : \mathbb{Z} \to \mathbb{R}$ such that

$$f(3x - y) \cdot f(y) = 3f(x)$$

holds for all integers x and y.

83. Determine all functions $f : \mathbb{R} \longrightarrow \mathbb{R}$ such that

$$f(xf(y) + x) = xy + f(x)$$

holds for all $x, y \in \mathbb{R}$.

84. Find all functions $f : \mathbb{R} \to \mathbb{R}$ such that the equation

$$f(f(x + y)) = f(x) + y$$

is satisfied for arbitrary $x, y \in \mathbb{R}$.

SOLUTIONS

Problem 80. (A-I-3-14) *Let $f : \mathbb{R} \to \mathbb{R}$ be a real function satisfying*

$$f(f(x) - y) = x - f(y) \quad \textit{for all } x, y \in \mathbb{R}.$$

Prove that f is an odd function, i.e., $f(-x) = -f(x)$ holds for all $x \in \mathbb{R}$.

Solution: Substituting $y = 0$ in the initial equation we obtain

$$f(f(x)) = x - f(0) \tag{1}$$

and thus

$$f(f(x)) - x = -f(0) \quad \text{for all } x \in \mathbb{R}. \tag{2}$$

If we put $x = f(y)$ into the given equation, then we get

$$f(f(f(y)) - y) = 0. \tag{3}$$

Using (2) in (3) we get $f(-f(0)) = 0$. Further, if we put $x = -f(0)$ into (1), we obtain

$$f(f(-f(0))) = -f(0) - f(0) = -2f(0).$$

The last equation yields $f(0) = -2f(0)$ and so $f(0) = 0$. Finally, putting $x = 0$ into the initial equation we get

$$f(-y) = f(f(0) - y) = -f(y) \quad \text{for all } y \in \mathbb{R}.$$

This means that f is an odd function and the proof is complete. □

Problem 81. (A-I-2-13) *Let* $\mathbb{R}^+ = (0; +\infty)$. *Determine all functions* $f : \mathbb{R}^+ \to \mathbb{R}$, *such that*

$$x\,f(x) = x\,f\left(\frac{x}{y}\right) + y\,f(y)$$

holds for all positive real values of x and y.

Solution: Let t be an arbitrary positive real number. For $x = ty$ we obtain

$$ty\,f(ty) = ty\,f(t) + y\,f(y).$$

Rewriting this equation we get

$$f(ty) = f(t) + \frac{f(y)}{t}.$$

for arbitrary positive t and y. Exchanging t and y we have

$$f(yt) = f(y) + \frac{f(t)}{y}.$$

Comparing the right sides of the last two equations, we get

$$f(t)\left(1-\frac{1}{y}\right)=f(y)\left(1-\frac{1}{t}\right).$$

Substituting $y=2$ with the notation $a=2f(2)\in\mathbb{R}$, we obtain

$$f(t)=2f(2)\left(1-\frac{1}{t}\right)=a\left(1-\frac{1}{t}\right).$$

After some easy checking we can see that the function $f(x)=a\left(1-\frac{1}{x}\right)$ satisfies the given equation for arbitrary real a. □

Problem 82. (A-T-3-08) *Determine all functions* $f:\mathbb{Z}\to\mathbb{R}$ *such that*

$$f(3x-y)\cdot f(y)=3f(x)$$

holds for all integers x *and* y.

Solution: Let f be any such function. Setting $x=y=0$ we get $f(0)\in\{0,3\}$.

If $f(0)=0$, we set $y=0$ and get $f(x)=0$, and so $f(x)\equiv 0$ is the first solution.

Now let $f(0)=3$. We set $x=0$ and get $f(y)f(-y)=9$, and particularly $f(y)\neq 0$ for all integers y. When we set $x=y\neq 0$ we get $f(2x)=3$. Thus $f(x)=3$ holds for all even values of x. Setting $y=4x$ gives us $f(-y)=f(y)$. Since $f(y)f(-y)=9$, it follows that $f(y)\in\{-3,3\}$ and f is an even function.

Setting $x=1$ and $y=2t+1$ (where t is an integer), we get

$$f(2-2t)\cdot f(2t+1)=3f(1),$$

from which it follows that $f(2t+1)=f(1)$ must hold. Checking, we see that $f(x)\equiv 3$ is the second solution and

$$f(x)=\begin{cases}3 & \text{for } x \text{ even},\\ -3 & \text{for } x \text{ odd}\end{cases}$$

is the third solution.

In conclusion, we see that the three functions

$$f(x) \equiv 0; \quad f(x) \equiv 3 \quad \text{and} \quad f(x) = \begin{cases} 3 & \text{for } x \text{ even,} \\ -3 & \text{for } x \text{ odd} \end{cases}$$

satisfy the given functional equation in the set of all integers. □

Problem 83. (A-T-1-04) *Determine all functions* $f : \mathbb{R} \longrightarrow \mathbb{R}$ *such that*

$$f(xf(y) + x) = xy + f(x)$$

holds for all $x, y \in \mathbb{R}$.

Solution: If we first assume any fixed value $x \neq 0$, simple manipulation gives us

$$\frac{f(xf(y) + x) - f(x)}{x} = y.$$

for all $y \in \mathbb{R}$.

We can now prove that f is an injective function. Assume, to the contrary, then for some $y_1 \neq y_2$ we have $f(y_1) = f(y_2)$. Then substitution in the left-hand side of the above equation gives us the same value $f(y_1)$ and $f(y_2)$, even though the right-hand sides are different. This contradiction ends the proof of the injectivity of the function f.

Setting $y = 0$ in the equation we have

$$f[xf(0) + x] = f(x), \quad \text{and therefore} \quad f[x(f(0) + 1)] = f(x),$$

and from the injectivity of the function f we therefore have $f(0) + 1 = 1$, which gives us $f(0) = 0$.

Setting $x = y = 1$, we next obtain

$$f[f(1) + 1] = 1 + f(1),$$

and this shows us that the number $a := f(1) + 1$ is a fixed point of the function f.

Let us now consider two possible cases:

Case 1: Assume $a = 0$. Then for $y = 1$ we have $f[xf(1) + x] = x + f(x)$, and therefore

$$f[x(f(1) + 1)] = x + f(x),$$

and for $a = f(1) + 1 = 0$ we get $f(0) = x + f(x)$. Because $f(0) = 0$, this yields $f(x) = -x$. It can easily be checked that the function $f(x) = -x$ satisfies the given functional equation for every $x, y \in \mathbb{R}$.

Case 2: Assume $a \neq 0$. Setting $x = a$, we obtain $f[af(y)+a] = ay+a$. From the right-hand side of this expression, we see that for $y \in \mathbb{R}$ the set of values of the function f is equal to the entire set $\mathbb{R}$. There must therefore exist $b \in \mathbb{R}$ such that $f(b) = -1$.

Setting $y = b$, we have

$$f[xf(b) + x] = xb + f(x).$$

Because $f(b) = -1$, we obtain $f(-x + x) = xb + f(x)$ or $f(0) = bx + f(x)$ and since $f(0) = 0$, we have $f(x) = -bx$.

Since we now have $f(x) = -bx$, the left-hand side of the equation gives us

$$f[xf(y) + x] = -b[xf(y) + x] = -b[-bxy + x] = b^2xy - bx,$$

and the right-hand side gives us

$$xy + f(x) = xy - bx.$$

Comparing these, we obtain $b^2 = 1$, i.e., $b = -1$ or $b = 1$.

The case $b = -1$ was considered before. If $b = 1$, it can be easily checked that the function $f(x) = x$ also satisfies the given equation for every $x, y \in \mathbb{R}$.

We see that the given functional equation has exactly two solutions, namely $f(x) = -x$ and $f(x) = x$ defined for all $x \in \mathbb{R}$. □

Problem 84. (A-I-2-01) *Find all functions* $f : \mathbb{R} \to \mathbb{R}$ *such that the equation*

$$f(f(x + y)) = f(x) + y$$

is satisfied for arbitrary $x, y \in \mathbb{R}$.

Solution: Exchanging x and y, we obtain

$$f(f(y+x)) = f(y) + x \qquad \text{for all } x, y \in \mathbb{R}.$$

From the functional equation, it immediately follows that

$$f(x) + y = f(y) + x \qquad \text{for all } x, y \in \mathbb{R}.$$

Taking $y = 0$ we get

$$f(x) = x + f(0).$$

For the left side of the given equation, we get the expression

$$\begin{aligned} f(f(x+y)) &= f(x+y+f(0)) \\ &= (x+y+f(0)) + f(0) = x + y + 2f(0) \end{aligned}$$

and for its the right side we have

$$f(x) + y = (x + f(0)) + y = x + y + f(0).$$

From last two conditions we get $f(0) = 0$ and therefore $f(x) = x$ for all $x \in \mathbb{R}$. Furthermore, we see (after checking) that the unique solution of the given equation is therefore the function $f(x) = x$ for all $x \in \mathbb{R}$.

The given functional equation thus has the unique solution $f(x) = x$.

□

3.4. Polynomials

Problems on polynomials can be very closely related to functional equations or inequalities. Still, the type of reasoning required in this area can be quite distinctive.

PROBLEMS

85. Let V be a real function defined by the expression

$$V(x) = (x-1)(x-2) + (x-1)(x-2)(x-3)(x-4) + (x-3)(x-4).$$

(a) Determine the minimum value of $V(x)$.
(b) Determine all values of x for which this minimum value is assumed.

86. Prove that the polynomial

$$P_n(x) = 3(2x-1)^{2n+2} - 3x^{2n+2} - 9x^2 + 12x - 3$$

is divisible by the polynomial

$$Q(x) = 3x^3 - 4x^2 + x$$

for all positive integers n.

87. Determine all polynomials $P(x)$ such that

$$P^2(x) = P(xy)P\left(\frac{x}{y}\right)$$

holds for all real numbers x, $y \neq 0$.

88. Determine all polynomials $P(x)$ with real coefficients, such that for each real number x the equation

$$P(x^2) - 3x^3 + 15x^2 - 24x + 12 = P(x)P(2x)$$

holds.

SOLUTIONS

Problem 85. (A-T-3-05) *Let V be a real function defined by the expression*

$$V(x) = (x-1)(x-2) + (x-1)(x-2)(x-3)(x-4) + (x-3)(x-4).$$

(a) *Determine the minimum value of $V(x)$.*
(b) *Determine all values of x for which this minimum value is assumed.*

Solution: First of all, we will consider a solution to this problem using calculus. We can write the given polynomial in the form

$$V(x) = (x^2 - 3x + 2) + (x^2 - 3x + 2) \cdot (x^2 - 7x + 12) + (x^2 - 7x + 12).$$

We calculate

$$V'(x) = 2x - 3 + (2x - 3) \cdot (x^2 - 7x + 12) + (x^2 - 3x + 2) \cdot (2x - 7) + 2x - 7,$$

which simplifies to

$$V'(x) = 4x^3 - 30x^2 + 74x - 60.$$

The necessary condition for the existence of an extreme value is a zero of the derivative. This means that we must solve the equation

$$4x^3 - 30x^2 + 74x - 60 = 0.$$

Noting that

$$V'(2) = 2 \cdot 8 - 30 \cdot 4 + 74 \cdot 2 - 60 = 0$$

holds, we have a potential extreme value in $x_1 = 2$. In order to completely solve the third degree algebraic equation, we must divide the polynomial $V'(x) = 4x^3 - 30x^2 + 74x - 60$ by $x - 2$. From this division, we obtain the equation

$$4x^2 - 22x + 30 = 0 \iff 2x^2 - 11x + 15 = 0.$$

The roots of this quadratic equation are $x_2 = \frac{5}{2}$ and $x_3 = 3$. We can now check the sufficient condition for the existence of an extreme value. In order to do this, we calculate

$$V''(x) = 12x^2 - 60x + 74,$$

and this shows us

$$V''(2) = 2 > 0, \quad V''(\tfrac{5}{2}) = -1 < 0, \quad \text{and} \quad V''(3) = 2 > 0.$$

We see that $x_1 = 2$ and $x_3 = 3$ are local minima of the function V and $x_2 = \frac{5}{2}$ is a local maximum.

We can now calculate $V(2) = 2$ and $V(3) = 2$. Since both values are the same, we see that the function assumes its global minimum in two points, namely $x_1 = 2$ and $x_3 = 3$, and the minimal value in these points is $V_{\min} = 2$. □

Another solution: A solution not using calculus, that might be considered a bit more elegant, is the following. The building blocks of the polynomial V are linear polynomials with the roots 1, 2, 3 and 4. These are clearly symmetric with respect to 2.5. This suggests utilising a substitution of $x + 2.5$ for x. This substitution gives us the new polynomial

$$\begin{aligned} U(x) = V(x + 2.5) &= \left(x + \tfrac{3}{2}\right)\left(x + \tfrac{1}{2}\right) \\ &\quad + \left(x + \tfrac{3}{2}\right)\left(x + \tfrac{1}{2}\right)\left(x - \tfrac{3}{2}\right)\left(x - \tfrac{1}{2}\right) + \left(x - \tfrac{3}{2}\right)\left(x - \tfrac{1}{2}\right) \\ &= x^4 - \tfrac{1}{2}x^2 + \tfrac{33}{16}. \end{aligned}$$

The new polynomial U is a even function. It therefore suffices to consider only values $x \geq 0$. Putting $x^2 = t$ in the polynomial U, we obtain the somewhat simpler function

$$Z(t) = t^2 - \tfrac{1}{2}t + \tfrac{33}{16} = (t - \tfrac{1}{4})^2 + \tfrac{32}{16} \quad \text{for } t \geq 0.$$

The function Z takes its minimum value for $t = \frac{1}{4}$, which translates to $x_1 = -\frac{1}{2}$ and $x_2 = \frac{1}{2}$ for the points in which U assumes its minimum values. Translation by 2.5 therefore gives us the points in which the original polynomial V assumes its minimum values, and these are therefore $x_1 = 2$ and $x_2 = 3$. This minimal value is easily calculated as

$$Z(\tfrac{1}{4}) = \tfrac{1}{16} - \tfrac{1}{2} \cdot \tfrac{1}{4} + \tfrac{33}{16} = \tfrac{34}{16} - \tfrac{1}{8} = 2. \qquad \square$$

Problem 86. (A-I-3-03) *Prove that the polynomial*

$$P_n(x) = 3(2x-1)^{2n+2} - 3x^{2n+2} - 9x^2 + 12x - 3$$

is divisible by the polynomial

$$Q(x) = 3x^3 - 4x^2 + x$$

for all positive integers n.

Solution: The proof is by induction on n. First, we check that the theorem is true for $n = 0$. Since

$$\begin{aligned} P_0(x) &= 3(2x-1)^2 - 3x^2 - 9x^2 + 12x - 3 \\ &= 3(4x^2 - 4x + 1) - 12x^2 + 12x - 3 = 0, \end{aligned}$$

this is obviously true, and we have the base for our induction.

We now assume that $Q(x)|P_n(x)$ for some specific value of $n \in N$. We aim to prove that $Q(x)|P_{n+1}(x)$ then also holds. Let us consider

$$P_{n+1}(x) = 3(2x-1)^{2n+4} - 3x^{2n+4} - 9x^2 + 12x - 3.$$

By adding and subtracting the same expression, we get

$$\begin{aligned} P_{n+1}(x) &= 3(2x-1)^{2n+4} - 3x^{2n+4} - 3(2x-1)^{2n+2} + 3x^{2n+2} \\ &\quad + 3(2x-1)^{2n+2} - 3x^{2n+2} - 9x^2 + 12x - 3 = W(x) + P_n(x), \end{aligned}$$

where

$$\begin{aligned} W(x) &= 3(2x-1)^{2n+4} - 3x^{2n+4} - 3(2x-1)^{2n+2} + 3x^{2n+2} \\ &= 3(2x-1)^{2n+2}[(2x-1)^2 - 1] + 3x^{2n+2}(1 - x^2) \end{aligned}$$

$$= 3(2x-1)^{2n+2}(4x^2-4x) + 3x^{2n+2}(1-x^2)$$
$$= 12(2x-1)^{2n+2}x(x-1) + 3x^{2n+2}(1-x^2).$$

The polynomial $Q(x)$ divides the polynomial $P_n(x)$ by our induction hypothesis. In order to complete the induction, we must show that it also divides the polynomial $W(x)$. In order to show this, it is sufficient to show that all roots of the polynomial $Q(x)$ are also roots of the polynomial $W(x)$. We have

$$Q(x) = 3x^3 - 4x^2 + x = x(3x^2 - 4x + 1) = x(x-1)(3x-1),$$

and it follows that the roots of Q are $x_1 = 0$, $x_2 = 1$ and $x_3 = \frac{1}{3}$.

It now remains for us to check that these roots of $Q(x)$ are also roots of $W(x)$. It is quite straightforward to check that

$$W(0) = 0, \quad W(1) = 0, \quad W(\tfrac{1}{3}) = 0$$

indeed hold, and the proof is then complete. □

Problem 87. (A-T-1-99) *Determine all polynomials $P(x)$ such that*

$$P^2(x) = P(xy)P\left(\frac{x}{y}\right)$$

holds for all real numbers x, $y \neq 0$.

Solution: Let $n \in N \cup \{0\}$ be the degree of the polynomial P, and a_i be the coefficients of each x^i for $0 \leq i \leq n$.

We first set $y = x$ for $x \neq 0$. This gives us the equation $P^2(x) = P(x^2) \cdot P(1)$.

Let us denote $c = P(1)$ and consider two possible cases:

Case 1: Assume $c = 0$. In this case, we have $P^2(x) = 0$, and therefore $P(x) = 0$ for every $x \neq 0$. We can easily check that $P(x) \equiv 0$ indeed fulfils the given equation for $x \in \mathbb{R}$ and we have therefore found the first solution $P(x) \equiv 0$.

Case 2: Assume $c \neq 0$. Then we have the equation $P^2(x) = P(x^2) \cdot c$. We can write the polynomial P in the form $P(x) = a_n x^n + a_j x^j + Q(x)$, where $a_n \neq 0$, the polynomial is of degree n and j is the second highest power of x for which the coefficient in P is not equal to 0. (Q is a polynomial of degree less than j.) We then have

$$P^2(x) = a_n^2 x^{2n} + 2a_j a_n x^{n+j} + W(x),$$

where $W(x)$ is a polynomial of degree less than $n + j$. Furthermore, we also have

$$c \cdot P(x^2) = c \cdot a_n x^{2n} + c \cdot a_j x^{2j} + T(x),$$

where $T(x)$ is a polynomial of degree less than $2j$. These polynomials must be equal. Since the coefficient of x^{n+j} in the second expression is equal to 0 (as $2j < j + n$ certainly holds), the equation $2a_j \cdot a_n = 0$ must hold. Since $a_n \neq 0$, this implies $a_j = 0$. We see that the polynomial P must be a monomial of degree n, i.e., $P(x) = a_n x^n$. We must now check that every such polynomial indeed fulfils the given equation:

$$P^2(x) = \left(a_n x^n\right)^2 = a_n^2 x^{2n}$$

and

$$P(xy) \cdot P\left(\frac{x}{y}\right) = a_n (xy)^n \cdot a_n \left(\frac{x}{y}\right)^n = a_n^2 x^{2n}.$$

We have therefore obtained the second solution $P_2(x) = a_n x^n$ for all $n \in N \cup \{0\}$ (which, we note, includes the first solution as a special case).

In summary, we see that the solutions of the given functional equation are all monomials $P(x) = c \cdot x^n$, $x \in R$ for all $n \in N \cup \{0\}$ and arbitrary real c. □

Problem 88. (A-T-3-16) *Determine all polynomials $P(x)$ with real coefficients, such that for each real number x the equation*

$$P(x^2) - 3x^3 + 15x^2 - 24x + 12 = P(x)P(2x)$$

holds.

Solution: It is easy to see that the polynomial $P(x)$ has degree equal to at least 2.

Let n, k be integers with $n \geq 2$ and $n > k \geq 0$. We can write the polynomial $P(x)$ in the form

$$P(x) = a_n x^n + a_k x^k + a_{k-1} x^{k-1} + \cdots + a_1 x + a_0,$$

where the a_i for $i \in \{n, k, k-1, \dots, 1, 0\}$ are real coefficients and $a_n \neq 0$. Then we have

$$P(x^2) = a_n x^{2n} + a_k x^{2k} + a_{k-1} x^{2(k-1)} + \cdots + a_1 x^2 + a_0,$$
$$P(2x) = 2^n a_n x^n + 2^k a_k x^k + 2^{k-1} a_{k-1} x^{k-1} + \cdots + 2a_1 x + a_0.$$

Comparing coefficients at x^{2n} on both sides of the equation, we obtain $a_n = 2^n a_n^2$ and from $a_n \neq 0$ it follows that $a_n = (\frac{1}{2})^n$ must hold.

The highest exponent of x on the left-hand side of the equation is equal to $2n$ and the second highest exponent is equal to $\max\{2k, 3\}$. The second-highest exponent of x on the right-hand side is $n + k$.

Since $n + k > 2k$ we can see that for $a_k \neq 0$ we have $n + k = 3$, and therefore $n = 2$ and $k = 1$ or $n = 3$ and $k = 0$.

In the case $n = 3$ and $k = 0$, the desired polynomial $P(x)$ is of the form $P(x) = \frac{1}{8}x^3 + A$, where A is a real number. Substituting in the original equation we have

$$\tfrac{1}{8}x^6 + A - 3x^3 + 15x^2 - 24x + 12 = (\tfrac{1}{8}x^3 + A)(x^3 + A) = \tfrac{1}{8}x^6 + \tfrac{9}{8}x^3 + A^2.$$

Comparing coefficients at x^2, we obtain $15 = 0$, and so this case is impossible.

In the case $n = 2$ and $k = 1$, the desired polynomial $P(x)$ is of the form $P(x) = \frac{1}{4}x^2 + Ax + B$, where A and B are real numbers. Substituting in the original equation we have

$$\begin{aligned}
&\tfrac{1}{4}x^4 + Ax^2 + B - 3x^3 + 15x^2 - 24x + 12 \\
&= (\tfrac{1}{4}x^2 + Ax + B)(x^2 + 2Ax + B) \\
&= \tfrac{1}{4}x^4 + \tfrac{3}{2}Ax^3 + (\tfrac{5}{4}B + 2A^2)x^2 + 3ABx + B^2.
\end{aligned}$$

Comparing coefficients for x^i, $i \in \{3, 2, 1, 0\}$ we obtain

$$-3 = \tfrac{3}{2}A, \quad A + 15 = \tfrac{5}{4}B + 2A^2, \quad -24 = 3AB, \quad B + 12 = B^2,$$

and solving this system yields $A = -2$, $B = 4$.

There therefore exists a unique polynomial $P(x)$ satisfying the given equation, namely $P(x) = \frac{1}{4}x^2 - 2x + 4$. □

3.5. Inequalities

Inequalities are a staple of many olympiad-style competitions, and the Mathematical Duel is no exception. This section includes some of the more interesting inequality problems posed over the years.

PROBLEMS

89. Prove that the inequality

$$(a+9)\left(a^2+\frac{1}{a}+\frac{b^2}{8}\right) \geq (a+b+1)^2$$

holds for all positive real values a and b. When does equality hold?

90. We are given positive real numbers x, y, z, u with $xyzu = 1$. Prove

$$\frac{x^3}{y^3}+\frac{y^3}{z^3}+\frac{z^3}{u^3}+\frac{u^3}{x^3} \geq x^2+y^2+z^2+u^2.$$

When does equality hold?

91. Let $a_1, a_2, \ldots, a_5$ be positive real numbers. Prove

$$\frac{a_1+a_2}{a_3+a_4+a_5}+\frac{a_2+a_3}{a_4+a_5+a_1}+\frac{a_3+a_4}{a_5+a_1+a_2}$$
$$+\frac{a_4+a_5}{a_1+a_2+a_3}+\frac{a_5+a_1}{a_2+a_3+a_4} \geq \frac{10}{3}.$$

When does equality hold?

92. Let a, b, c be arbitrary positive real numbers such that $abc = 1$. Prove that the inequality

$$\frac{a}{ab+1}+\frac{b}{bc+1}+\frac{c}{ca+1} \geq \frac{3}{2}$$

holds. When does equality hold?

93. Let a, b, c be arbitrary positive real numbers. Prove that the inequality

$$\left(1+\frac{a}{b+c}\right)\left(1+\frac{b}{c+a}\right)\left(1+\frac{c}{a+b}\right) \geq \frac{27}{8}$$

holds. When does equality hold?

94. Let a, b, c be non-zero real numbers. Prove that the inequality

$$\frac{a^2}{b^2}+\frac{b^2}{c^2}+\frac{c^2}{a^2}\geq\frac{a}{b}+\frac{b}{c}+\frac{c}{a}$$

holds. When does equality hold?

95. Let a, b, c be arbitrary positive real numbers. Prove that the inequality

$$\frac{1}{b(a+b)}+\frac{1}{c(b+c)}+\frac{1}{a(c+a)}\geq\frac{9}{2(a^2+b^2+c^2)}$$

holds. When does equality hold?

96. Let a and b be positive real numbers such that $a^2+b^2=1$ holds. Prove

$$\frac{a}{b^2+1}+\frac{b}{a^2+1}\geq\frac{2}{3}\cdot(a+b).$$

When does equality hold?

97. Prove that the inequality

$$\frac{1}{ab(a+b)}+\frac{1}{bc(b+c)}+\frac{1}{ca(c+a)}\geq\frac{9}{2(a^3+b^3+c^3)}$$

is satisfied for any positive real numbers a, b, c. When does equality hold?

98. Prove that the inequality

$$\frac{ab}{cd}+\frac{bc}{da}+\frac{cd}{ab}+\frac{da}{bc}\geq\frac{a}{c}+\frac{b}{d}+\frac{c}{a}+\frac{d}{b}$$

is satisfied for all positive real numbers a, b, c, d. When does equality hold?

99. Prove that the inequality

$$a(a+b)+b(b+1)\geq-\frac{1}{3}$$

holds for all real numbers a and b. When does equality hold?

100. Let a, b, c be arbitrary real numbers. Prove that the inequality

$$a^2+5b^2+4c^2\geq 4(ab+bc)$$

holds. When does equality hold?

SOLUTIONS

Problem 89. (A-T-1-14) *Prove that the inequality*

$$(a+9)\left(a^2+\frac{1}{a}+\frac{b^2}{8}\right) \geq (a+b+1)^2$$

holds for all positive real values a and b. When does equality hold?

Solution: We first note that applying the Cauchy inequality to the triples

$$\left(\frac{u}{\sqrt{x}}, \frac{v}{\sqrt{y}}, \frac{w}{\sqrt{z}}\right) \quad \text{and} \quad (\sqrt{x}, \sqrt{y}, \sqrt{z})$$

yields

$$(u+v+w)^2 \leq \left(\frac{u^2}{x}+\frac{v^2}{y}+\frac{w^2}{z}\right)(x+y+z)$$

or

$$\frac{(u+v+w)^2}{x+y+z} \leq \frac{u^2}{x}+\frac{v^2}{y}+\frac{w^2}{z}.$$

The given inequality is equivalent to

$$\frac{(a+b+1)^2}{1+8+a} \leq \frac{a^2}{1}+\frac{b^2}{8}+\frac{1}{a},$$

which is exactly the inequality derived from the Cauchy inequality, setting $u=a$, $v=b$, $w=1$, $x=1$, $y=8$ and $z=a$.

Equality holds if and only if the triples are multiples of each other, i.e. if

$$\frac{u/\sqrt{x}}{\sqrt{x}} = \frac{u}{x} = \frac{v}{y} = \frac{w}{z} \quad \text{or} \quad \frac{a}{1} = \frac{b}{8} = \frac{1}{a}$$

holds. Since $a = \frac{1}{a}$ and $a > 0$ hold, we have $a = 1$ and therefore $b = 8$ as the only case in which equality holds. □

Problem 90. (A-T-2-13) *We are given positive real numbers* x, y, z, u *with* $xyzu = 1$. *Prove*

$$\frac{x^3}{y^3}+\frac{y^3}{z^3}+\frac{z^3}{u^3}+\frac{u^3}{x^3} \geq x^2+y^2+z^2+u^2.$$

When does equality hold?

Solution: Applying the arithmetic–geometric mean inequality for the six positive numbers

$$\frac{x^3}{y^3}, \frac{x^3}{y^3}, \frac{x^3}{y^3}, \frac{y^3}{z^3}, \frac{y^3}{z^3}, \frac{z^3}{u^3}$$

we have

$$\frac{1}{6}\left(3 \cdot \frac{x^3}{y^3} + 2 \cdot \frac{y^3}{z^3} + \frac{z^3}{u^3}\right) \geq \sqrt[6]{\frac{x^9}{y^3 z^3 u^3}} = \sqrt[6]{x^{12}} = x^2.$$

Cyclically, we also obtain

$$\frac{1}{6}\left(3 \cdot \frac{y^3}{z^3} + 2 \cdot \frac{z^3}{u^3} + \frac{u^3}{x^3}\right) \geq y^2,$$

$$\frac{1}{6}\left(3 \cdot \frac{z^3}{u^3} + 2 \cdot \frac{u^3}{x^3} + \frac{x^3}{y^3}\right) \geq z^2,$$

$$\frac{1}{6}\left(3 \cdot \frac{u^3}{x^3} + 2 \cdot \frac{x^3}{y^3} + \frac{y^3}{z^3}\right) \geq u^2.$$

Adding all the above four inequalities we obtain the required inequality.

By the arithmetic–geometric mean inequality, equality holds if and only if all the terms are equal. This means

$$\left(\frac{x}{y}\right)^3 = \left(\frac{y}{z}\right)^3 = \left(\frac{z}{u}\right)^3, \quad \text{or} \quad \frac{x}{y} = \frac{y}{z} = \frac{z}{u}.$$

Similarly, the next three inequalities yield

$$\frac{y}{z} = \frac{z}{u} = \frac{u}{x}, \quad \frac{x}{y} = \frac{z}{u} = \frac{u}{x} \quad \text{and} \quad \frac{u}{x} = \frac{x}{y} = \frac{y}{z}.$$

Taken together, these equalities yield

$$\frac{x}{y} = \frac{y}{z} = \frac{z}{u} = \frac{u}{x}.$$

We therefore obtain

$$\left(\frac{x}{y}\right)^4 = \frac{x}{y} \cdot \frac{y}{z} \cdot \frac{z}{u} \cdot \frac{u}{x} = 1,$$

from the assumption $xyzu = 1$, which yields

$$\left(\frac{x}{y}\right)^4 = 1, \quad \text{and therefore} \quad \frac{x}{y} = 1.$$

Since all these fractions are equal, we therefore have

$$\frac{x}{y} = \frac{y}{z} = \frac{z}{u} = \frac{u}{x} = 1,$$

and therefore $x = y = z = u$. Because of $xyzu = 1$, this gives us equality if and only if $x = y = z = u = 1$. □

Another solution: For the proof, we can also apply the rearrangement inequality for the quadruples

$$\left(\sqrt{\frac{x^3}{y^3}}, \sqrt{\frac{y^3}{z^3}}, \sqrt{\frac{z^3}{u^3}}, \sqrt{\frac{u^3}{x^3}}\right), \left(\frac{x}{y}, \frac{y}{z}, \frac{z}{u}, \frac{u}{x}\right), \left(\sqrt{\frac{x}{y}}, \sqrt{\frac{y}{z}}, \sqrt{\frac{z}{u}}, \sqrt{\frac{u}{x}}\right).$$

Problem 91. (A-T-1-09) *Let $a_1, a_2, \ldots, a_5$ be positive real numbers. Prove*

$$\frac{a_1 + a_2}{a_3 + a_4 + a_5} + \frac{a_2 + a_3}{a_4 + a_5 + a_1} + \frac{a_3 + a_4}{a_5 + a_1 + a_2}$$
$$+ \frac{a_4 + a_5}{a_1 + a_2 + a_3} + \frac{a_5 + a_1}{a_2 + a_3 + a_4} \geq \frac{10}{3}.$$

When does equality hold?

Solution: We define five positive numbers:

$$x_i = \frac{a_i + a_{i+1} + a_{i+2}}{a_1 + a_2 + a_3 + a_4 + a_5} \quad \text{for } i = 1, 2, 3, 4, 5, \quad \text{where } a_{i+5} = a_i.$$

We have

$$\frac{1}{x_1} + \frac{1}{x_2} + \frac{1}{x_3} + \frac{1}{x_4} + \frac{1}{x_5}$$
$$= 5 + \frac{a_1 + a_2}{a_3 + a_4 + a_5} + \frac{a_2 + a_3}{a_4 + a_5 + a_1}$$
$$+ \frac{a_3 + a_4}{a_5 + a_1 + a_2} + \frac{a_4 + a_5}{a_1 + a_2 + a_3} + \frac{a_5 + a_1}{a_2 + a_3 + a_4}.$$

Therefore, the harmonic mean of the numbers x_i can be written as

$$H(x_1, x_2, x_3, x_4, x_5)$$
$$= \frac{5}{\frac{1}{x_1} + \frac{1}{x_2} + \frac{1}{x_3} + \frac{1}{x_4} + \frac{1}{x_5}}$$
$$= \frac{5}{5 + \frac{a_1+a_2}{a_3+a_4+a_5} + \frac{a_2+a_3}{a_4+a_5+a_1} + \frac{a_3+a_4}{a_5+a_1+a_2} + \frac{a_4+a_5}{a_1+a_2+a_3} + \frac{a_5+a_1}{a_2+a_3+a_4}}$$

Since

$$x_1 + x_2 + x_3 + x_4 + x_5 = 3,$$

the arithmetic mean of the numbers x_i can be written as

$$A(x_1, x_2, x_3, x_4, x_5) = \frac{x_1 + x_2 + x_3 + x_4 + x_5}{5} = \frac{3}{5}.$$

By the harmonic–arithmetic means inequality

$$H(x_1, x_2, x_3, x_4, x_5) \le A(x_1, x_2, x_3, x_4, x_5),$$

we therefore obtain

$$\frac{5}{5 + \frac{a_1+a_2}{a_3+a_4+a_5} + \frac{a_2+a_3}{a_4+a_5+a_1} + \frac{a_3+a_4}{a_5+a_1+a_2} + \frac{a_4+a_5}{a_1+a_2+a_3} + \frac{a_5+a_1}{a_2+a_3+a_4}} \le \frac{3}{5},$$

which implies

$$\frac{a_1 + a_2}{a_3 + a_4 + a_5} + \frac{a_2 + a_3}{a_4 + a_5 + a_1} + \frac{a_3 + a_4}{a_5 + a_1 + a_2} + \frac{a_4 + a_5}{a_1 + a_2 + a_3} + \frac{a_5 + a_1}{a_2 + a_3 + a_4} \ge \frac{10}{3}.$$

This completes the proof of the given inequality.

By the harmonic–arithmetic mean inequality, equality holds if and only if all the terms x_1, x_2, x_3, x_4, x_5 are equal.

Let S denote the sum $S = a_1 + a_2 + a_3 + a_4 + a_5$. By the definition of x_i for $i = 1, 2, 3, 4, 5$ we have the equalities

$$\frac{a_1 + a_2 + a_3}{S} = \frac{a_2 + a_3 + a_4}{S} = \frac{a_3 + a_4 + a_5}{S} = \frac{a_4 + a_5 + a_1}{S} = \frac{a_5 + a_1 + a_2}{S},$$

and consequently

$$a_1 + a_2 + a_3 = a_2 + a_3 + a_4 = a_3 + a_4 + a_5 = a_4 + a_5 + a_1 = a_5 + a_1 + a_2.$$

Subtracting successive sums, we obtain the equalities $a_1 - a_4 = 0$, $a_2 - a_5 = 0$, $a_3 - a_1 = 0$, $a_4 - a_2 = 0$, and consequently $a_1 = a_2 = a_3 = a_4 = a_5$. □

Problem 92. (A-I-3-08) *Let a, b, c be arbitrary positive real numbers such that $abc = 1$. Prove that the inequality*

$$\frac{a}{ab+1} + \frac{b}{bc+1} + \frac{c}{ca+1} \geq \frac{3}{2}$$

holds. When does equality hold?

Solution: Since $abc = 1$, there exist positive real numbers x, y, z such that

$$a = \frac{x}{y}, \qquad b = \frac{y}{z}, \qquad c = \frac{z}{x}.$$

Substitution gives us

$$\frac{a}{ab+1} + \frac{b}{bc+1} + \frac{c}{ca+1} = \frac{\frac{x}{y}}{\frac{x}{z}+1} + \frac{\frac{y}{z}}{\frac{y}{x}+1} + \frac{\frac{z}{x}}{\frac{z}{y}+1}$$

$$= \frac{zx}{xy+yz} + \frac{xy}{zx+yz} + \frac{yz}{xy+zx} \geq \frac{3}{2}.$$

This inequality follows from *Nesbitt's* inequality, which states

$$\frac{r}{s+t} + \frac{s}{t+r} + \frac{t}{r+s} \geq \frac{3}{2}$$

for positive real values of r, s, and t.

Equality holds if and only if $r = s = t$. From this, we obtain $a = b = c$. Because $abc = 1$, we therefore have $a = b = c = 1$. □

Problem 93. (A-I-1-06) *Let a, b, c be arbitrary positive real numbers. Prove that the inequality*

$$\left(1 + \frac{a}{b+c}\right)\left(1 + \frac{b}{c+a}\right)\left(1 + \frac{c}{a+b}\right) \geq \frac{27}{8}$$

holds. When does equality hold?

Solution: We define three positive real numbers

$$x = b + c, \quad y = a + c, \quad z = a + b.$$

The arithmetic mean of these numbers is given by the expression

$$A(x, y, z) = \frac{x+y+z}{3} = \frac{2}{3}(a+b+c),$$

and the geometric mean by

$$G(x, y, z) = \sqrt[3]{xyz} = \sqrt[3]{(b+c)(a+c)(a+b)}.$$

By the arithmetic–geometric mean inequality

$$A(x, y, z) \geq G(x, y, z),$$

we therefore obtain

$$\tfrac{2}{3}(a+b+c) \geq \sqrt[3]{(b+c)(a+c)(a+b)}.$$

This is equivalent to

$$\tfrac{8}{27}(a+b+c)^3 \geq (a+b)(b+c)(c+a)$$

and therefore

$$\left(1+\frac{a}{b+c}\right)\left(1+\frac{b}{c+a}\right)\left(1+\frac{c}{a+b}\right) \geq \frac{27}{8}$$

follows, which completes the proof of the inequality.

Equality holds if and only if $x = y = z$, which is equivalent to $b + c = a + c = a + b$. Subtraction yields $b - a = 0$ and $c - b = 0$ and therefore $a = b = c$. □

Problem 94. (A-I-2-05) *Let a, b, c be non-zero real numbers. Prove that the inequality*

$$\frac{a^2}{b^2} + \frac{b^2}{c^2} + \frac{c^2}{a^2} \geq \frac{a}{b} + \frac{b}{c} + \frac{c}{a}$$

holds. When does equality hold?

Solution: First we prove the given inequality for positive numbers a, b, c. We define

$$x := \frac{a}{b}, \qquad y := \frac{b}{c}, \qquad z := \frac{c}{a}.$$

Then x, y, z are also positive numbers with

$$xyz = \frac{a}{b} \cdot \frac{b}{c} \cdot \frac{c}{a} = 1.$$

Writing the arithmetic mean and geometric mean in the form

$$A(x, y, z) = \frac{x+y+z}{3}, \qquad G(x, y, z) = \sqrt[3]{xyz},$$

the arithmetic–geometric means inequality gives us

$$A(x, y, z) \geq G(x, y, z).$$

We note that $G(x, y, z) = 1$ holds, because $xyz = 1$.

We therefore have

$$A(x, y, z) \geq 1, \quad \text{and thus} \quad A^2(x, y, z) \geq A(x, y, z).$$

In the last expression, equality holds if and only if $A(x, y, z) = 1$.

From the definition of $A(x, y, z)$ we have

$$\left(\frac{x+y+z}{3}\right)^2 \geq \frac{x+y+z}{3}$$

and therefore

$$\frac{x^2+y^2+z^2+2xy+2xz+2yz}{9} \geq \frac{x+y+z}{3}.$$

Next, we can apply the inequality

$$x^2+y^2+z^2 \geq xy+xz+yz,$$

for which equality holds if and only if $x = y = z$. We then obtain

$$\frac{x^2+y^2+z^2+2x^2+2y^2+2z^2}{9} \geq \frac{x+y+z}{3},$$

which yields the given inequality for positive values of a, b, c.

Since the inequality holds for the absolute values of any real values of a, b, c, it is also true for all real values, independent of their sign. This concludes the proof of the given inequality.

We know that equality holds if and only if $A(x, y, z) = \frac{x+y+z}{3} = 1$, i.e., $x + y + z = 3$. Furthermore, equality holds for the second inequality if and only if $x = y = z$ holds. This means that equality holds if and only if $x = y = z = 1$. Returning to the original variables a, b, c we then obtain

$$\frac{a}{b} = \frac{b}{c} = \frac{c}{a} = 1,$$

which is equivalent to $a = b = c$. □

Problem 95. (A-I-1-04) *Let a, b, c be arbitrary positive real numbers. Prove that the inequality*

$$\frac{1}{b(a+b)}+\frac{1}{c(b+c)}+\frac{1}{a(c+a)} \geq \frac{9}{2(a^2+b^2+c^2)}$$

holds. When does equality hold?

Solution: We define the numbers

$$x=\frac{1}{b(a+b)}, \quad y=\frac{1}{c(b+c)}, \quad z=\frac{1}{a(c+a)},$$

noting that x, y and z are certainly positive.

We can write the arithmetic mean of these numbers as

$$A(x, y, z)=\frac{\frac{1}{b(a+b)}+\frac{1}{c(b+c)}+\frac{1}{a(c+a)}}{3},$$

and their harmonic mean as

$$H(x, y, z)=\frac{3}{\frac{1}{x}+\frac{1}{y}+\frac{1}{z}}=\frac{3}{b(a+b)+c(b+c)+a(c+a)}.$$

By the arithmetic–harmonic means inequality $A(x, y, z) \geq H(x, y, z)$, we obtain

$$\frac{\frac{1}{b(a+b)}+\frac{1}{c(b+c)}+\frac{1}{a(c+a)}}{3} \geq \frac{3}{b(a+b)+c(b+c)+a(c+a)},$$

or

$$\frac{1}{b(a+b)}+\frac{1}{c(b+c)}+\frac{1}{a(c+a)} \geq \frac{9}{b(a+b)+c(b+c)+a(c+a)}. \quad (1)$$

By the well-known inequality $a^2+b^2+c^2 \geq ab+bc+ca$, we have

$$\frac{9}{a^2+b^2+c^2+ab+bc+ca} \geq \frac{9}{2(a^2+b^2+c^2)}, \quad (2)$$

and combining (1) and (2) gives us the desired inequality, completing our proof.

Equality in (2) holds if and only if $a^2+b^2+c^2=ab+bc+ca$, which is true only for $a=b=c$. In the arithmetic–harmonic means inequality, equality holds if and only if $x=y=z$, i.e.,

$$\frac{1}{b(a+b)}=\frac{1}{c(b+c)}=\frac{1}{a(c+a)}.$$

We see that this condition is also fulfilled if $a=b=c$. □

Problem 96. (A-I-3-02) *Let a and b be positive real numbers such that $a^2+b^2=1$ holds. Prove*

$$\frac{a}{b^2+1}+\frac{b}{a^2+1}\ge\frac{2}{3}\cdot(a+b).$$

When does equality hold?

Solution: The given inequality can be written in the form

$$\frac{a^3+a+b^3+b}{(b^2+1)(a^2+1)}\ge\frac{2}{3}\cdot(a+b),$$

which is equivalent to

$$\frac{(a+b)\cdot(a^2-ab+b^2+1)}{a^2b^2+a^2+b^2+1}\ge\frac{2}{3}\cdot(a+b).$$

Dividing both sides by $(a+b)>0$ we see that the given inequality is equivalent to

$$\frac{a^2-ab+b^2+1}{a^2b^2+a^2+b^2+1}\ge\frac{2}{3}.$$

Using the condition $a^2+b^2=1$ we get the further equivalent inequality

$$\frac{2-ab}{a^2b^2+2}\ge\frac{2}{3}.$$

Substituting $x=ab$, we get the inequality

$$\frac{2-x}{x^2+2}\ge\frac{2}{3}, \tag{1}$$

which we must now prove for positive values of x.

Furthermore, since $(a-b)^2\ge 0$, we have $a^2-2ab+b^2\ge 0$, and because $a^2+b^2=1$, we obtain $2ab\le 1$, or $x=ab\le\frac{1}{2}$.

We can therefore assume that $0<x\le\frac{1}{2}$.

The inequality (1) is equivalent to

$$6-3x\ge 2(x^2+2)\iff 2x^2+3x-2\le 0.$$

The quadratic equation $2x^2+3x-2=0$ has the roots $x_1=-2$ and $x_2=\frac{1}{2}$. It therefore follows that (1) certainly holds for $0<x\le\frac{1}{2}$, which completes the proof of the given inequality.

Equality holds if and only if $x=\frac{1}{2}$. This is equivalent to $ab=\frac{1}{2}$, and because we have $a^2+b^2=1$, this implies $a^2-2ab+b^2=0$ or

$(a-b)^2 = 0$ and therefore $a = b$. From $a^2 + b^2 = 1$ we therefore obtain $a = b = \frac{\sqrt{2}}{2}$. □

Problem 97. (A-T-1-01) *Prove that the inequality*

$$\frac{1}{ab(a+b)} + \frac{1}{bc(b+c)} + \frac{1}{ca(c+a)} \geq \frac{9}{2(a^3+b^3+c^3)}$$

is satisfied for any positive real numbers a, b, c. When does equality hold?

Solution: We consider the inequality

$$a^3 + b^3 \geq ab(a+b) = a^2b + ab^2$$

which is true for arbitrary positive real numbers a, b. We rewrite this inequality in the form

$$\frac{1}{ab(a+b)} \geq \frac{1}{a^3+b^3}.$$

Likewise,

$$\frac{1}{bc(b+c)} \geq \frac{1}{b^3+c^3} \quad \text{and} \quad \frac{1}{ca(c+a)} \geq \frac{1}{c^3+a^3}$$

for all positive real numbers a, b, c.

Adding these last three inequalities yields

$$\frac{1}{ab(a+b)} + \frac{1}{bc(b+c)} + \frac{1}{ca(c+a)} \geq \frac{1}{a^3+b^3} + \frac{1}{b^3+c^3} + \frac{1}{c^3+a^3}.$$

Applying the arithmetic–harmonic mean inequality to the right side of the last inequality, we get

$$\frac{1}{3}\left(\frac{1}{a^3+b^3} + \frac{1}{b^3+c^3} + \frac{1}{c^3+a^3}\right) \geq \frac{3}{(a^3+b^3)+(b^3+c^3)+(c^3+a^3)},$$

and this gives us

$$\frac{1}{a^3+b^3} + \frac{1}{b^3+c^3} + \frac{1}{c^3+a^3} \geq \frac{9}{2(a^3+b^3+c^3)}.$$

By the arithmetic–harmonic mean inequality, equality holds if and only if $a = b = c$. Also the three first equalities are fulfilled if $a = b = c$. Therefore equality holds in the given inequality if and only if $a = b = c$ holds. □

Problem 98. (A-I-3-00) *Prove that the inequality*

$$\frac{ab}{cd}+\frac{bc}{da}+\frac{cd}{ab}+\frac{da}{bc}\geq\frac{a}{c}+\frac{b}{d}+\frac{c}{a}+\frac{d}{b}$$

is satisfied for all positive real numbers a, b, c, d. When does equality hold?

Solution: We define $x = a/c$ and $y = b/d$. By assumption, we certainly have $x, y > 0$. We can then write

$$\frac{1}{x}=\frac{c}{a}\quad\text{and}\quad\frac{1}{y}=\frac{d}{b}.$$

We will now prove the following lemma.

Lemma 9. *If x and y are positive real numbers, the inequality*

$$xy+\frac{y}{x}+\frac{1}{xy}+\frac{x}{y}\geq x+y+\frac{1}{x}+\frac{1}{y}$$

holds.

Proof. Applying the Cauchy inequality to the vectors $(x, 1)$ and $(1, y)$ gives us

$$\sqrt{x^2+1}\cdot\sqrt{y^2+1}\geq x+y,$$

and similarly applying the Cauchy inequality to the vectors $(x, 1)$ and $(y, 1)$ gives us

$$\sqrt{x^2+1}\cdot\sqrt{y^2+1}\geq xy+1.$$

Multiplying these inequalities yields

$$(x^2+1)(y^2+1)\geq(x+y)(xy+1),$$

which is equivalent to

$$x^2y^2+y^2+1+x^2\geq x^2y+xy^2+y+x,$$

and dividing this inequality by xy yields the required result. □

Continuing with our proof we now note that

$$xy + \frac{y}{x} + \frac{1}{xy} + \frac{x}{y} \geq x + y + \frac{1}{x} + \frac{1}{y}$$

is equivalent to

$$xy + \frac{y}{x} + \frac{1}{xy} + \frac{x}{y} - x - y - \frac{1}{x} - \frac{1}{y} \geq 0.$$

Adding 1 to both sides of this inequality yields

$$xy + \frac{y}{x} + \frac{1}{xy} + \frac{x}{y} - x - y - \frac{1}{x} - \frac{1}{y} + 1 \geq 1,$$

which can be written as

$$x\left(y + \frac{1}{y} - 1\right) + \frac{1}{x}\left(y + \frac{1}{y} - 1\right) - 1\left(y + \frac{1}{y} - 1\right) \geq 1$$

or

$$\left(x + \frac{1}{x} - 1\right) \cdot \left(y + \frac{1}{y} - 1\right) \geq 1.$$

By the well-known inequality $x + \frac{1}{x} \geq 2$ for $x > 0$, we have $x + \frac{1}{x} - 1 \geq 1$ and $y + \frac{1}{y} - 1 \geq 1$, and this inequality therefore certainly holds.

Equality holds if and only if $x + \frac{1}{x} - 1 = 1$ and $y + \frac{1}{y} - 1 = 1$, which is equivalent to $x + \frac{1}{x} = 2$ and $y + \frac{1}{y} = 2$ for $x > 0$ and $y > 0$. Equality therefore holds if and only if $x = y = 1$.

For the given variables, this holds if and only if $\frac{a}{c} = 1$ and $\frac{b}{d} = 1$ hold for $a, b, c, d > 0$. This is the case for $a = c$ and $b = d$ for $a, b, c, d > 0$. □

Problem 99. (A-I-3-99) *Prove that the inequality*

$$a(a + b) + b(b + 1) \geq -\tfrac{1}{3}$$

holds for all real numbers a and b. When does equality hold?

Solution: First of all, we rewrite the given inequality in the equivalent form $a^2 + ab + b^2 + b + \frac{1}{3} \geq 0$. We can interpret this as a quadratic inequality with the variable a and the parameter b.

We now consider the discriminant

$$D = \frac{b^2}{4} - b^2 - b - \frac{1}{3} = -\frac{3}{4}\left(b^2 + \frac{4}{3}\cdot b + \frac{4}{9}\right) = -\frac{3}{4}\left(b + \frac{2}{3}\right)^2$$

of the quadratic function $f(a) = a^2 + ab + b^2 + b + \frac{1}{3}$. This expression is certainly never positive for any real value of b, and the inequality therefore certainly holds.

Equality holds for $D = 0$, which is equivalent to $b = -\frac{2}{3}$ and the double root in this case is $a = \frac{-b}{2}$, yielding $a = \frac{1}{3}$ and $b = -\frac{2}{3}$. □

Problem 100. (B-I-1-16) *Let a, b, c be arbitrary real numbers. Prove that the inequality*

$$a^2 + 5b^2 + 4c^2 \geq 4(ab + bc)$$

holds. When does equality hold?

Solution: We can rewrite the given inequality in the equivalent form

$$(a - 2b)^2 + (b - 2c)^2 \geq 0.$$

Thus the given inequality holds for all real numbers a, b, c.

Equality holds if and only if $a = 2b$ and simultaneously $b = 2c$, i.e., for the triples $(a, b, c) = (4c, 2c, c)$ with arbitrary real number c. □

3.6. Algebra and Numbers

Some nominally algebraic problems really just deal with number properties. Here are a few such problems.

PROBLEMS

101. Let a, b, p, q and $p\sqrt{a} + q\sqrt{b}$ be positive rational numbers. Prove that the numbers $\sqrt{a}$ and $\sqrt{b}$ are also rational.

102. Let the equality

$$a^2 + b^2 + c^2 + d^2 + ad - cd = 1$$

be fulfilled for some real numbers a, b, c and d. Prove that $ab + bc$ cannot be equal to 1.

103. Determine all real numbers a for which the equations

$$x^4 + ax^2 + 1 = 0 \quad \text{and} \quad x^3 + ax + 1 = 0$$

have a common root.

104. Evaluate the following súms in terms of n:

(a) $\sum_{k=1}^{n} k! \cdot (k^2 + 1)$,
(b) $\sum_{k=1}^{n} 2^{n-k} \cdot k \cdot (k+1)!$.

SOLUTIONS

Problem 101. (A-I-3-11) *Let a, b, p, q and $p\sqrt{a}+q\sqrt{b}$ be positive rational numbers. Prove that the numbers $\sqrt{a}$ and $\sqrt{b}$ are also rational.*

Solution: Let us observe that $p\sqrt{a} + q\sqrt{b} > 0$ and

$$p^2a - q^2b = (p\sqrt{a} + q\sqrt{b})(p\sqrt{a} - q\sqrt{b}).$$

We can therefore write

$$p\sqrt{a} - q\sqrt{b} = \frac{p^2a - q^2b}{p\sqrt{a} + q\sqrt{b}}.$$

Since both the numerator and the denominator of the fraction are rational, so is the number $p\sqrt{a} - q\sqrt{b}$.

The rationality of $\sqrt{a}$ and $\sqrt{b}$ now follows from the following two equalities:

$$\sqrt{a} = \frac{(p\sqrt{a} + q\sqrt{b}) + (p\sqrt{a} - q\sqrt{b})}{2p},$$

$$\sqrt{b} = \frac{(p\sqrt{a} + q\sqrt{b}) - (p\sqrt{a} - q\sqrt{b})}{2q}$$

and the rationality of the numbers p, q, $p\sqrt{a} + q\sqrt{b}$ and $p\sqrt{a} - q\sqrt{b}$.

□

Problem 102. (A-I-1-02) *Let the equality*

$$a^2 + b^2 + c^2 + d^2 + ad - cd = 1$$

be fulfilled for some real numbers a, b, c *and* d. *Prove that* $ab + bc$ *cannot be equal to* 1.

Solution: Assume, to the contrary, that there exist real numbers a, b, c, d such that both $a^2 + b^2 + c^2 + d^2 + ad - cd = 1$ and $ab + bc = 1$ hold.

Subtracting $ab + bc = 1$ from the given equation, we obtain

$$a^2 + b^2 + c^2 + d^2 + ad - cd - ab - bc = 0,$$

which is equivalent to

$$2a^2 + 2b^2 + 2c^2 + 2d^2 + 2ad - 2cd - 2ab - 2bc = 0.$$

Rewriting this as

$$(a^2 - 2ab + b^2) + (b^2 - 2bc + c^2) + (c^2 - 2cd + d^2) + (a^2 + 2ad + d^2) = 0,$$

yields

$$(a - b)^2 + (b - c)^2 + (c - d)^2 + (a + d)^2 = 0.$$

This means that

$$(a - b)^2 = 0, \quad (b - c)^2 = 0, \quad (c - d)^2 = 0, \quad (a + d)^2 = 0,$$

must hold, and therefore

$$a - b = 0, \quad b - c = 0, \quad c - d = 0, \quad a + d = 0.$$

From the three first equations we obtain $a = b = c = d$, and using the fourth we get $a = b = c = d = 0$.

This contradicts our assumption that $ab + bc = 1$ holds. We see that $ab + bc$ cannot be equal to 1, as claimed. □

Problem 103. (A-I-1-00) *Determine all real numbers* a *for which the equations*

$$x^4 + ax^2 + 1 = 0 \quad \textit{and} \quad x^3 + ax + 1 = 0$$

have a common root.

Solution: Let x_0 be the common root of the two equations. We note that $x_0 \neq 0$ obviously holds.

The root x_0 fulfil both equations

$$x_0^4 + ax_0^2 + 1 = 0 \quad \text{and} \quad x_0^3 + ax_0 + 1 = 0.$$

Multiplying the second equation by $x_0 \neq 0$ and subtracting the result from the first equation gives us $1 - x_0 = 0$, or $x_0 = 1$. If a common root of the two given equations exists, we see that it must be equal to 1.

Inserting $x_0 = 1$ in both equations we twice obtain the equation $1 + a + 1 = 0$, and therefore conclude that $a = -2$ must hold. □

Problem 104. (A-T-1-02) *Evaluate the following sums in terms of n:*

(a) $\sum_{k=1}^{n} k! \cdot (k^2 + 1)$,

(b) $\sum_{k=1}^{n} 2^{n-k} \cdot k \cdot (k + 1)!$.

Solution: (a) We first consider the kth term of the sum:

$$\begin{aligned}(k^2 + 1)k! &= (k^2 + 1 + k - k)k! \\ &= (k^2 + k)k! - (k - 1)k! = k(k + 1)! - (k - 1)k!\end{aligned}$$

This gives us

$$\begin{aligned}\sum_{k=1}^{n}(k^2 + 1)k! &= \sum_{k=1}^{n} k(k + 1)! - \sum_{k=1}^{n}(k - 1)k! \\ &= \sum_{k=2}^{n+1}(k - 1)k! - \sum_{k=1}^{n}(k - 1)k! \\ &= n(n + 1)! - 0 \cdot 1 = n(n + 1)!\end{aligned}$$

(b) We can calculate the kth term of the sum in the following manner:

$$k(k + 1)! = [(k + 2) - 2](k + 1)! = (k + 2)! - 2(k + 1)!$$

and therefore

$$2^{n-k}k(k + 1)! = 2^{n-k}(k + 2)! - 2^{n-k+1}(k + 1)!$$

This gives us

$$\sum_{k=1}^{n} 2^{n-k} k(k+1)! = \sum_{k=1}^{n} 2^{n-k}(k+2)! - \sum_{k=1}^{n} 2^{n-(k-1)}(k+1)!$$

$$= \sum_{k=2}^{n+1} 2^{n-(k-1)}(k+1)! - \sum_{k=1}^{n} 2^{n-(k-1)}(k+1)!$$

$$= 2^{n-(n+1-1)}(n+2)! - 2^n \cdot 2! = (n+2)! - 2^{n+1}.$$

□

3.7. Trigonometry

Some problems in the Mathematical Duel deal with properties of trigonometric functions. The arguments required to solve such problems are mostly algebraic in nature, although there is always some geometric flavour in the background. Of course, it helps to know some of the more important formulas relating to trigonometric functions and their mutual relationships.

PROBLEMS

105. Let α, β, γ be the interior angles of a triangle with $\gamma > 90°$. Prove that the inequality

$$\tan\alpha \tan\beta < 1$$

holds.

106. Determine the minimum value of the expression

$$V = \frac{\sin\alpha}{\sin\beta\sin\gamma} + \frac{\sin\beta}{\sin\gamma\sin\alpha} + \frac{\sin\gamma}{\sin\alpha\sin\beta},$$

where α, β, γ are interior angles of a triangle.

107. Let α and β be the interior angles of a triangle ABC at A and B respectively, such that

$$1 + \cos^2(\alpha+\beta) = \cos^2\alpha + \cos^2\beta$$

holds. Prove that ABC is right angled.

108. In the coordinate plane, determine the set of all points with coordinates (x, y), such that the real numbers x, y satisfy the inequality

$$\sin x \geq \cos y,$$

with $x, y \in [-4, 4]$.

109. Let α, β, γ be the interior angles of a triangle. Prove that

$$\sin\alpha - \sin\beta + \sin\gamma = 4\sin\frac{\alpha}{2}\cos\frac{\beta}{2}\sin\frac{\gamma}{2}$$

holds.

110. Let $\alpha, \beta, \gamma, \delta$ be angles from the interval $[-\frac{\pi}{2}, \frac{\pi}{2}]$ such that

$$\sin\alpha + \sin\beta + \sin\gamma + \sin\delta = 1, \quad \text{and}$$
$$\cos 2\alpha + \cos 2\beta + \cos 2\gamma + \cos 2\delta \geq \frac{10}{3}$$

hold. Prove that $\alpha, \beta, \gamma, \delta \in [0, \frac{\pi}{6}]$ must hold.

SOLUTIONS

Problem 105. (A-I-4-13) *Let α, β, γ be the interior angles of a triangle with $\gamma > 90°$. Prove that the inequality*

$$\tan\alpha\tan\beta < 1$$

holds.

Solution: First of all, we note that $\alpha + \beta = 180° - \gamma < 90°$ holds, and therefore $\tan(\alpha + \beta) > 0$. We can now consider the well-known formula

$$\tan(\alpha + \beta) = \frac{\tan\alpha + \tan\beta}{1 - \tan\alpha \cdot \tan\beta}.$$

Since the numerator of the fraction on the right side is obviously positive, the denominator of the fraction must be positive as well. It therefore follows that

$$1 - \tan\alpha \cdot \tan\beta > 0 \iff \tan\alpha\tan\beta < 1$$

holds, and the proof is complete. □

Another solution: For $\alpha + \beta > 90°$ we certainly have $\cos(\alpha + \beta) > 0$. Because of

$$\cos(\alpha + \beta) = \cos\alpha \cos\beta - \sin\alpha \sin\beta > 0,$$

we have

$$\cos\alpha \cos\beta > \sin\alpha \sin\beta.$$

Since $\cos\alpha \cos\beta > 0$, dividing by the product of the cosines gives us

$$\tan\alpha \tan\beta = \frac{\sin\alpha}{\cos\alpha} \cdot \frac{\sin\beta}{\cos\beta} < 1,$$

which proves the given inequality. □

Another solution: Let CP denote the altitude from C and h its length. Let D be the point on the side AB, such that $\angle ACD = 90°$. Let $|AP| = p$, $|PD| = q$ and $|DB| = r$ (see Fig. (a)). We then have

$$\tan\alpha \cdot \tan\beta = \frac{h}{p} \cdot \frac{h}{q+r} < \frac{h}{p} \cdot \frac{h}{q} = \frac{h^2}{pq} = 1,$$

which completes the proof. □

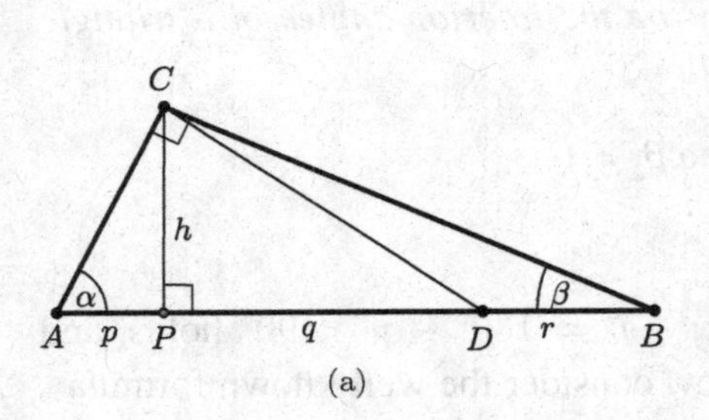

(a)

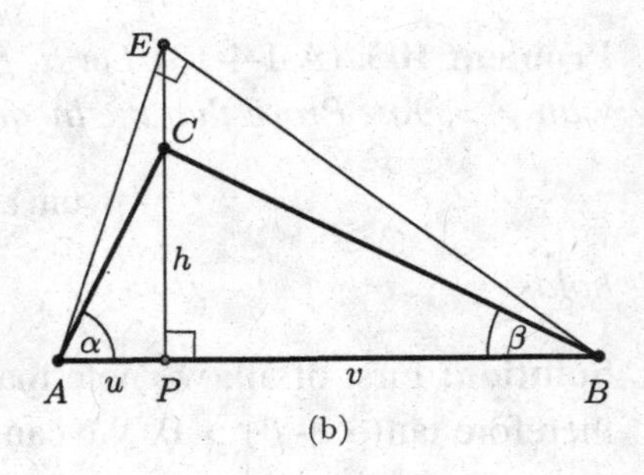

(b)

Another solution: As in the previous solution we consider the altitude CP of the length h. On the ray PC we choose the point E, such that ABE is a right-angled triangle with hypotenuse AB (see Fig. (b)). Finally, we write $|PE| = w > h$, $|AP| = u$ and $|PB| = v$. We then obtain the following:

$$\tan\alpha \cdot \tan\beta = \frac{h}{u} \cdot \frac{h}{v} < \frac{w}{u} \cdot \frac{w}{v} = \frac{w^2}{uv} = 1,$$

and the proof is finished. □

Problem 106. (A-I-4-12) *Determine the minimum value of the expression*

$$V = \frac{\sin\alpha}{\sin\beta\sin\gamma} + \frac{\sin\beta}{\sin\gamma\sin\alpha} + \frac{\sin\gamma}{\sin\alpha\sin\beta},$$

where α, β, γ are interior angles of a triangle.

Solution: Since α, β, γ are interior angles of a triangle, the values of $\sin\alpha$, $\sin\beta$ and $\sin\gamma$ are positive real numbers. We can therefore apply the AM–GM inequality in the form

$$V = \frac{\sin\alpha}{\sin\beta\sin\gamma} + \frac{\sin\beta}{\sin\gamma\sin\alpha} + \frac{\sin\gamma}{\sin\alpha\sin\beta} \geq 3\sqrt[3]{\frac{1}{\sin\alpha\sin\beta\sin\gamma}}.$$

Since the function $\sin x$ is concave on $(0;\pi)$, we can estimate the denominator of the right-hand side of the last inequality by Jensen's inequality (combined with the AM–GM inequality) in the following way:

$$\begin{aligned}\sqrt[3]{\sin\alpha\sin\beta\sin\gamma} &\leq \frac{\sin\alpha + \sin\beta + \sin\gamma}{3}\\ &\leq \sin\left(\frac{\alpha+\beta+\gamma}{3}\right) = \sin\frac{\pi}{3} = \frac{\sqrt{3}}{2}.\end{aligned}$$

Thus, we have $V \geq 2\sqrt{3}$ with equality for $\alpha = \beta = \gamma = \frac{1}{3}\pi$, i.e., in the case of the equilateral triangle.

The minimum value of the expression V is therefore $2\sqrt{3}$. □

Another solution: Let a, b, c be the lengths of the sides and P the area of the triangle. Rewriting the expression V and applying the law of sines, we obtain

$$\begin{aligned}V &= \frac{\sin^2\alpha}{\sin\alpha\sin\beta\sin\gamma} + \frac{\sin^2\beta}{\sin\alpha\sin\beta\sin\gamma} + \frac{\sin^2\gamma}{\sin\alpha\sin\beta\sin\gamma}\\ &= \frac{a^2}{bc\sin\alpha} + \frac{b^2}{ca\sin\beta} + \frac{c^2}{ab\sin\gamma}.\end{aligned}$$

Since

$$bc\sin\alpha = ca\sin\beta = ac\sin\gamma = 2P,$$

we therefore have

$$V = \frac{a^2+b^2+c^2}{2P}.$$

By the well-known inequality

$$a^2 + b^2 + c^2 \geq 4\sqrt{3}P$$

(see for example Problem 2 of the third IMO), we obtain the required inequality $v \geq 2\sqrt{3}$. □

Another solution: Using the AM–GM inequality, we have

$$\frac{\sin\alpha}{\sin\beta\sin\gamma} + \frac{\sin\beta}{\sin\gamma\sin\alpha} \geq \frac{2}{\sin\gamma},$$
$$\frac{\sin\beta}{\sin\alpha\sin\gamma} + \frac{\sin\gamma}{\sin\alpha\sin\beta} \geq \frac{2}{\sin\alpha},$$
$$\frac{\sin\gamma}{\sin\alpha\sin\beta} + \frac{\sin\alpha}{\sin\beta\sin\gamma} \geq \frac{2}{\sin\beta}.$$

Adding all three inequalities and using Jensen's inequality for the convex function $\frac{1}{\sin x}$ on $(0;\pi)$, we obtain

$$V \geq \frac{1}{\sin\alpha} + \frac{1}{\sin\beta} + \frac{1}{\sin\gamma} \geq \frac{3}{\sin\left(\frac{\alpha+\beta+\gamma}{3}\right)} = 2\sqrt{3},$$

as required. □

Problem 107. (A-I-3-09) *Let α and β be the interior angles of a triangle ABC at A and B respectively, such that*

$$1 + \cos^2(\alpha+\beta) = \cos^2\alpha + \cos^2\beta$$

holds. Prove that ABC is right angled.

Solution: We can rewrite the given property in the form

$$1 + 1 - \sin^2(\alpha+\beta) = 1 - \sin^2\alpha + 1 - \sin^2\beta$$

or

$$\sin^2(\alpha+\beta) = \sin^2\alpha + \sin^2\beta.$$

In a triangle, the angles certainly fulfil the property $\alpha + \beta = 180° - \gamma$, and we therefore have

$$\sin^2(\pi - \gamma) = \sin^2\alpha + \sin^2\beta,$$

which we can write in the form

$$\sin^2 \gamma = \sin^2 \alpha + \sin^2 \beta.$$

Using the well-known formula $\sin \alpha = \frac{a}{2R}$, where R is the circumradius of the triangle, we obtain

$$\frac{c^2}{4R^2} = \frac{a^2}{4R^2} + \frac{b^2}{4R^2}.$$

From this, we see that $c^2 = a^2 + b^2$ must hold, and the Pythagorean theorem tells us that the triangle is right angled, with the right angle in C. □

Problem 108. (A-T-1-03) *In the coordinate plane, determine the set of all points with coordinates* (x, y), *such that the real numbers* x, y *satisfy the inequality*

$$\sin x \geq \cos y,$$

with $x, y \in [-4, 4]$.

Solution: Let S denote the square $[-4, 4] \times [-4, 4]$. We can write the given inequality in the form

$$\sin x - \sin\left(\frac{\pi}{2} - y\right) \geq 0,$$

and applying the trigonometric formula

$$\sin \alpha - \sin \beta = 2 \sin \frac{\alpha - \beta}{2} \cos \frac{\alpha + \beta}{2},$$

yields the equivalent inequality

$$2 \sin \frac{x + y - \frac{\pi}{2}}{2} \cos \frac{x - y + \frac{\pi}{2}}{2} \geq 0.$$

We will now examine the equations

$$\sin \frac{x + y - \frac{\pi}{2}}{2} = 0 \quad \text{and} \quad \cos \frac{x - y + \frac{\pi}{2}}{2} = 0.$$

For the equation

$$\sin \frac{x + y - \frac{\pi}{2}}{2} = 0,$$

the general solution is

$$\frac{x+y-\frac{\pi}{2}}{2}=k\cdot\pi,\ \text{ with } k\in\mathbb{Z}.$$

Because $(x, y) \in S$, we have $k \in \{-1, 0, 1\}$, yielding three relevant solutions for our purpose:

$$y=-x-\frac{3\pi}{2},\quad y=-x+\frac{\pi}{2},\quad y=-x+\frac{5\pi}{2}$$

(see Fig. (a)).

These three lines divide the square S into four parts. In each part the sign of the expression $\sin\frac{x+y-\frac{\pi}{2}}{2}$ is the same, i.e., either all negative or all positive. On each line, the expression is equal to zero. (Note that the graphs of the linear functions resulting from other values of k do not cross S.)

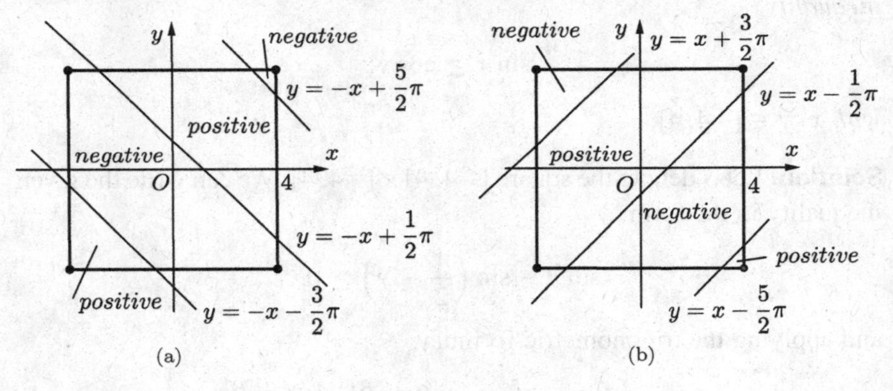

(a) (b)

For the equation

$$\cos\frac{x-y+\frac{\pi}{2}}{2}=0,$$

the general solution is

$$\frac{x-y+\frac{\pi}{2}}{2}=\frac{\pi}{2}+m\cdot\pi,\ \text{with } m\in\mathbb{Z}.$$

As before, because $(x, y) \in S$, we have $m \in \{-1, 0, 1\}$, again yielding three relevant solutions for our purpose:

$$y=x-\frac{\pi}{2},\quad y=x-\frac{5\pi}{2},\quad y=x+\frac{3\pi}{2}$$

(see Fig. (b)).

These three lines also divide S into four parts. In each part the sign of the expression $\cos\frac{x-y+\frac{\pi}{2}}{2}$ is the same, as was the case for the sine function. On the each line the expression is equal to the zero. (Once again, no other lines resulting from solutions cross S.)

Using the signs of the factors $\sin\frac{x+y-\frac{\pi}{2}}{2}$ and $\cos\frac{x-y+\frac{\pi}{2}}{2}$ we can determine the sign of the product in each section of the square S. The six relevant lines crossing S (three from Fig. (a) and three from Fig. (b)) divide the square S into ten parts. On all of these lines, the product is always equal to zero. The sign in any of the ten parts is uniformly positive or negative throughout. This result is shown in Fig. (c).

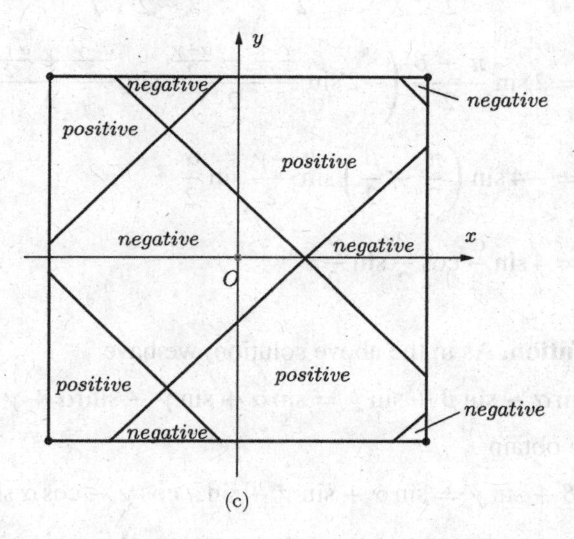

(c)

The solution of the inequality thus consists of the set of all points with coordinates (x, y), represented in the square S by areas labelled "positive", including all points on the borders of these areas. □

Problem 109. (A-I-4-95) *Let α, β, γ be the interior angles of a triangle. Prove that*

$$\sin\alpha - \sin\beta + \sin\gamma = 4\sin\frac{\alpha}{2}\cos\frac{\beta}{2}\sin\frac{\gamma}{2}$$

holds.

Solution: In a triangle, we certainly have $\beta = \pi - (\alpha + \gamma)$. We therefore obtain

$$\begin{aligned}
&\sin\alpha - \sin\beta + \sin\gamma \\
&\quad = \sin\alpha + \sin\gamma - \sin(\pi - (\alpha+\gamma)) \\
&\quad = \sin\alpha + \sin\gamma - \sin(\alpha+\gamma) \\
&\quad = 2\sin\frac{\alpha+\gamma}{2}\cos\frac{\alpha-\gamma}{2} - 2\sin\frac{\alpha+\gamma}{2}\cos\frac{\alpha+\gamma}{2} \\
&\quad = 2\sin\frac{\alpha+\gamma}{2}\left(\cos\frac{\alpha-\gamma}{2} - \cos\frac{\alpha+\gamma}{2}\right) \\
&\quad = 2\sin\frac{\pi-\beta}{2}\left(-2\sin\frac{\frac{\alpha-\gamma}{2} - \frac{\alpha+\gamma}{2}}{2}\sin\frac{\frac{\alpha-\gamma}{2} + \frac{\alpha+\gamma}{2}}{2}\right) \\
&\quad = -4\sin\left(\frac{\pi}{2} - \frac{\beta}{2}\right)\sin\frac{-\gamma}{2}\sin\frac{\alpha}{2} \\
&\quad = 4\sin\frac{\alpha}{2}\cos\frac{\beta}{2}\sin\frac{\gamma}{2}.
\end{aligned}$$

□

Another solution. As in the above solution, we have

$$\sin\alpha - \sin\beta + \sin\gamma = \sin\alpha + \sin\gamma - \sin(\alpha+\gamma).$$

We therefore obtain

$$\begin{aligned}
\sin\alpha - \sin\beta + \sin\gamma &= \sin\alpha + \sin\gamma - \sin\alpha\cos\gamma - \cos\alpha\sin\gamma \\
&= \sin\alpha(1 - \cos\gamma) + \sin\gamma(1 - \cos\alpha) \\
&= \sin\alpha \cdot 2\sin^2\frac{\gamma}{2} + \sin\gamma \cdot 2\sin^2\frac{\alpha}{2} \\
&= 2\left(\sin\alpha \cdot \sin^2\frac{\gamma}{2} + \sin\gamma \cdot \sin^2\frac{\alpha}{2}\right) \\
&= 2\left(2\sin\frac{\alpha}{2}\cos\frac{\alpha}{2}\sin^2\frac{\gamma}{2} + 2\sin\frac{\gamma}{2}\cos\frac{\gamma}{2}\sin^2\frac{\alpha}{2}\right) \\
&= 4\sin\frac{\alpha}{2}\sin\frac{\gamma}{2}\left(\cos\frac{\alpha}{2}\sin\frac{\gamma}{2} + \cos\frac{\gamma}{2}\sin\frac{\alpha}{2}\right) \\
&= 4\sin\frac{\alpha}{2}\sin\frac{\gamma}{2}\sin\left(\frac{\alpha}{2} + \frac{\gamma}{2}\right)
\end{aligned}$$

$$= 4\sin\frac{\alpha}{2}\sin\frac{\gamma}{2}\sin\frac{\alpha+\gamma}{2}$$

$$= 4\sin\frac{\alpha}{2}\sin\frac{\gamma}{2}\sin\frac{\pi-\beta}{2}$$

$$= 4\sin\frac{\alpha}{2}\cos\frac{\beta}{2}\sin\frac{\gamma}{2}.$$

□

Problem 110. (A-T-2-95) *Let* $\alpha, \beta, \gamma, \delta$ *be angles from the interval* $[-\frac{\pi}{2}, \frac{\pi}{2}]$ *such that*

$$\sin\alpha + \sin\beta + \sin\gamma + \sin\delta = 1, \text{ and}$$

$$\cos 2\alpha + \cos 2\beta + \cos 2\gamma + \cos 2\delta \geq \tfrac{10}{3}$$

hold. Prove that $\alpha, \beta, \gamma, \delta \in [0, \frac{\pi}{6}]$ *must hold.*

Solution: Using the formula $\cos 2x = 1 - 2\sin^2 x$, we can rewrite the given inequality in the form

$$1 - 2\sin^2\alpha + 1 - 2\sin^2\beta + 1 - 2\sin^2\gamma + 1 - 2\sin^2\delta \geq \tfrac{10}{3},$$

which simplifies to

$$-2\sin^2\alpha - 2\sin^2\beta - 2\sin^2\gamma - 2\sin^2\delta \geq -\tfrac{2}{3}$$

or

$$\sin^2\alpha + \sin^2\beta + \sin^2\gamma + \sin^2\delta \leq \tfrac{1}{3}.$$

In order to simplify notation, we define new variables $a = \sin\alpha$, $b = \sin\beta$, $c = \sin\gamma$ and $d = \sin\delta$. We can then write the assumptions of the problem in the form

$$a+b+c+d = 1, \quad \text{and} \quad a^2+b^2+c^2+d^2 \leq \tfrac{1}{3} \quad \text{with} \quad a, b, c, d \in [-1, 1].$$

We wish to prove $a, b, c, d \in [0, \frac{1}{2}]$.

Applying the root-mean-square–arithmetic-mean inequality, we obtain

$$\sqrt{\frac{a^2+b^2+c^2}{3}} \geq \frac{a+b+c}{3} = \frac{1-d}{3},$$

which gives us

$$\frac{a^2+b^2+c^2}{3} \geq \frac{(1-d)^2}{9}.$$

It follows that $a^2 + b^2 + c^2 \geq \frac{(1-d)^2}{3}$ holds, and adding d^2 to both sides of the inequality gives us

$$a^2 + b^2 + c^2 + d^2 \geq \tfrac{1}{3}\left(4d^2 - 2d + 1\right).$$

Let us now define the quadratic function

$$f(d) = \tfrac{1}{3}(4d^2 - 2d + 1)$$

on the segment $[-1, 1]$.

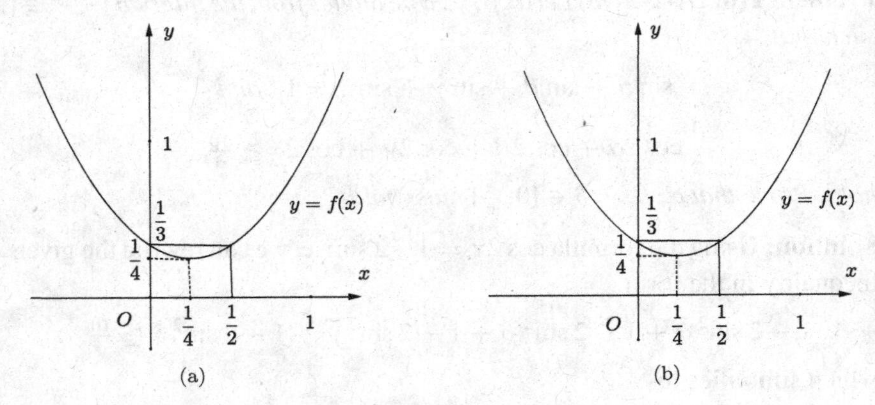

From the inequality we just derived, we know that $a^2+b^2+c^2+d^2 \geq f(d)$ for $d \in [-1, 1]$. We have $f(0) = f(\frac{1}{2}) = \frac{1}{3}$ and $f(\frac{1}{4}) = \frac{1}{4}$. We note (see Fig. (a)) that for $d > \frac{1}{2}$ we have $a^2 + b^2 + c^2 + d^2 > f(d) > \frac{1}{3}$. Similarly (see Fig. (b)), for $d < 0$ we also have $a^2 + b^2 + c^2 + d^2 > f(d) > \frac{1}{3}$.

This proves $0 \leq d \leq \frac{1}{2}$. The same argument can also be used to prove $0 \leq a, b, c \leq \frac{1}{2}$, completing the proof that $a, b, c, d \in [0, \frac{1}{2}]$. This means that $\alpha, \beta, \gamma, \delta \in [0, \frac{\pi}{6}]$, completing the proof. □

Chapter 4

Combinatorics

4.1. Counting Problems

In the parlance of modern mathematics competitions, the term "combinatorics" has come to mean something slightly more general than just the study of enumerative problems. Nowadays, the term includes questions relating to winning strategies for mathematical games, problems relating to graph theory or invariant theory, and more.

Still, even now the classic counting problem is a staple of mathematics competitions, and the Mathematical Duel is not an exception. In some problems, we count numbers, in some we count words and in some we count geometric figures with a specific property, but the answer is always a number.

PROBLEMS

111. A number is called *bumpy* if its digits alternately rise and fall from left to right. For instance, the number 36 180 is bumpy, because $3 < 6$, $6 > 1$, $1 < 8$ and $8 > 0$ all hold. On the other hand, neither 3 451 nor 81 818 are bumpy.

(a) Determine the difference between the largest and smallest five-digit bumpy numbers.
(b) How many five-digit bumpy numbers have the middle digit 5?
(c) Determine the total number of five-digit bumpy numbers.

112. A *Goodword* is a string of letters, in which there is always at least one vowel between any two consonants, i.e., in which no two consonants appear next to each other. We wish to form Goodwords from the letters of the word "duel".

(a) How many different four-letter Goodwords can be formed using all four letters?
(b) How many different four-letter Goodwords can be formed with these letters if they can each be used more than once (and therefore not all letters must be used in any specific Goodwords)?

113. A square $ABCD$ of the size 7×7 is divided into 49 smaller congruent squares using line segments parallel to its sides. Determine the total number of paths from the vertex A to the vertex C along lines of the resulting grid, if we are only allowed to move in the directions of the vectors $\overrightarrow{AB}$ and $\overrightarrow{AD}$.

114. One large cube $ABCDEFGH$ is formed from eight small congruent cubes. Determine the total number of paths from the vertex A to the vertex G along the edges of the small cubes, if we are only allowed to move in the directions of the vectors $\overrightarrow{AB}$, $\overrightarrow{AD}$ and $\overrightarrow{AE}$.

115. Points $P_1, P_2, \ldots, P_{2006}$ are placed at random intervals on a circle, such that no two of them are in the same spot. Determine the total number of different convex polygons, whose vertices are all among the points P_i.

116. Determine the number of triangles with integer-length sides, such that two of the sides are of length m and n $(1 \leq m \leq n)$. Solve this problem for the special values $m = 6$ and $n = 9$ and then for general values.

117. We are given two parallel lines p and q. Let us consider a set A of 13 different points such that 7 of them lie on p and the other 6 lie on q.

(a) How many line segments exist, whose endpoints are elements of the set A?
(b) How many triangles exist, whose vertices are elements of the set A?

118. Determine in how many ways one can assign numbers of the set $\{1, 2, \ldots, 8\}$ to the vertices of a cube $ABCDEFGH$ such that the sum of any two numbers at vertices with a common edge is always an odd number.

SOLUTIONS

Problem 111. (A-T-3-04) *A number is called bumpy if its digits alternately rise and fall from left to right. For instance, the number* 36 180 *is bumpy,* *because* $3 < 6$, $6 > 1$, $1 < 8$ *and* $8 > 0$ *all hold. On the other hand, neither* 3 451 *nor* 81 818 *are bumpy.*

(a) *Determine the difference between the largest and smallest five-digit bumpy numbers.*
(b) *How many five-digit bumpy numbers have the middle digit* 5?
(c) *Determine the total number of five-digit bumpy numbers.*

Solution: (a) The largest bumpy number is obviously 89 898 and the smallest 12 010. We therefore obtain $89\,898 - 12\,010 = 77\,888$ as the required difference.

(b) If the middle digit of a five-digit number $abcde$ is 5, the digits b and d can be 6, 7, 8 or 9. If d is 6, there are six possible values for e, namely 0, 1, 2, 3, 4, 5. similarly, there are seven values for $d = 7$, eight for $d = 8$ and nine for $d = 9$, giving a total of $6 + 7 + 8 + 9 = 30$ possible pairs for d and e. For each of the possible values of b, there is one less corresponding option for a, since $a = 0$ is not possible. This means that we have a total of $5 + 6 + 7 + 8 = 26$ possible pairs for a and b. Since these pairs are independent of each other, we obtain a total of $26 \cdot 30 = 780$ five-digit bumpy numbers with the middle digit 5.

(c) The argument for $c = 5$ is completely analogous for other possible values of c from 0 to 8. (Note that c cannot equal 9, as b and d must be greater than c.) We therefore obtain

$$(1+2+3+4+5+6+7+8+9)(1+2+3+4+5+6+7+8)$$
$$= 45 \cdot 36 = 1620 \text{ possibilities}$$

for $c = 0$,

$$(2+3+4+5+6+7+8+9)(1+2+3+4+5+6+7+8)$$
$$= 44 \cdot 36 = 1584$$

for $c = 1$,

$$(3+4+5+6+7+8+9)(2+3+4+5+6+7+8)$$
$$= 42 \cdot 35 = 1470$$

for $c = 2$,

$$(4+5+6+7+8+9)(3+4+5+6+7+8) = 39 \cdot 33 = 1287$$

for $c = 3$,

$$(5+6+7+8+9)(4+5+6+7+8) = 35 \cdot 30 = 1050$$

for $c = 4$,

$$(7+8+9)(6+7+8) = 24 \cdot 21 = 504$$

for $c = 6$,

$$(8+9)(7+8) = 17 \cdot 15 = 255$$

for $c = 7$

and

$$9 \cdot 8 = 72$$

for $c = 8$.

This yields a total of

$$1620 + 1584 + 1470 + 1287 + 1050 + 780 + 504 + 255 + 72 = 8622$$

possible bumpy five-digit numbers. □

Problem 112. (C-I-3-09) *A* Goodword *is a string of letters, in which there is always at least one vowel between any two consonants, i.e., in which no two consonants appear next to each other. We wish to form Goodwords from the letters of the word "duel".*

(a) *How many different 4-letter Goodwords can be formed using all four letters?*
(b) *How many different 4-letter Goodwords can be formed with these letters if they can each be used more than once (and therefore not all letters must be used in any specific Goodwords)?*

Solution: (a) Since the word "duel" contains two consonants and two vowels, we can note that there are three possible combinations of consonants (c) and vowels (v) that satisfy the conditions of the problem, namely cvvc, cvcv and vcvc. Each of these cases can be satisfied by placing the consonants in either of two orders and the vowels in either of two orders. There are therefore $2 \cdot 2 = 4$ Goodwords of each type, and therefore $3 \cdot 4 = 12$ such

Goodwords altogether. It is quite easy to list them all:

$$deul, duel, leud, lued, delu, dule, ledu, lude, edul, udel, elud, uled.$$

(b) Since both consonants need not appear, Goodwords can also be of the form *cvvv*, *vcvv*, *vvcv*, *vvvc* or *vvvv* in this case, along with the three types considered in (a). (Note that there cannot be 3 or 4 consonants, since two of these would then certainly appear next to another.) In each of the eight possible orders, each spot can be occupied by either of the two options (i.e., d or l if it must be a consonant and e or u if it must be a vowel). There are therefore $2^4 = 16$ Goodwords in each case, yielding a total of $8 \cdot 16 = 128$ four-letter Goodwords of this type. □

Problem 113. (C-T-2-06) *A square $ABCD$ of the size 7×7 is divided into 49 smaller congruent squares using line segments parallel to its sides. Determine the total number of paths from the vertex A to the vertex C along lines of the resulting grid, if we are only allowed to move in the directions of the vectors $\overrightarrow{AB}$ and $\overrightarrow{AD}$.*

Solution: Any path from A to C is made up of 14 small steps, each of which is the length of the sides of the small squares. Further, 7 of these are in the direction of $\overrightarrow{AB}$ and the other 7 are in the direction of $\overrightarrow{AD}$. Every possible path therefore results from a choice of 7 out of the 14 steps, all of which are in the direction of $\overrightarrow{AB}$ (since the other 7 are then automatically in the direction of $\overrightarrow{AD}$). This total number is therefore equal to $\binom{14}{7} = 3432$. □

Problem 114. (B-T-2-06) *One large cube $ABCDEFGH$ is formed from eight small congruent cubes. Determine the total number of paths from the vertex A to the vertex G along the edges of the small cubes, if we are only allowed to move in the directions of the vectors $\overrightarrow{AB}$, $\overrightarrow{AD}$ and $\overrightarrow{AE}$.*

Solution: As in the previous problem, each path is composed of six small steps. Two of these point in each of the allowed directions $\overrightarrow{AB}$, $\overrightarrow{AD}$ and $\overrightarrow{AE}$. The total number is therefore equal to the number of choices of 2 from these 6, which point in the direction of $\overrightarrow{AB}$, multiplied by the number of choices of two of the remaining four in the direction of $\overrightarrow{AD}$ (since the final two must then point in the direction of $\overrightarrow{AE}$). This total number of steps is therefore equal to $\binom{6}{2} \cdot \binom{4}{2} = 15 \cdot 6 = 90$. □

Problem 115. (A-T-3-06) *Points $P_1, P_2, \ldots, P_{2006}$ are placed at random intervals on a circle, such that no two of them are in the same spot. Determine the total number of different convex polygons, whose vertices are all among the points P_i.*

Solution: Any choice of n points out of these 2006 (with $3 \le n \le 2006$) yields a unique convex polygon. The number of polygons is therefore equal to the number of possible choices of points.

Since each point can either be chosen or not, the total number of subsets of the 2006-element set of points is therefore equal to 2^{2006}. The only ones of these subsets that cannot be the vertices of a polygon are those with less than three elements. There is 1 such set with no elements (the empty set). Furthermore, there are 2006 such subsets with exactly one element and $\binom{2006}{2} = 2011015$ such sets with exactly two elements. The total number of convex polygons is therefore equal to $2^{2006} - 2013022$. □

Problem 116. (C-I-1-08) *Determine the number of triangles with integer-length sides, such that two of the sides are of length m and n $(1 \le m \le n)$. Solve this problem for the special values $m = 6$ and $n = 9$ and then for general values.*

Solution: In the general case, the triangle inequality shows us that the length of the third side is greater than or equal to $n - m + 1$ and less than or equal to $n + m - 1$. There are $(n + m - 1) - (n - m + 1) + 1 = 2m - 1$ positive integers from $n + m - 1$ through $n - m + 1$, and the number of such triangles is therefore equal to $2m - 1$.

In the case $m = 6$ and $n = 9$, there are therefore 11 such triangles. □

Problem 117. (C-I-2-08) *We are given two parallel lines p and q. Let us consider a set A of 13 different points such that 7 of them lie on p and the other 6 lie on q.*

(a) *How many line segments exist, whose endpoints are elements of the set A?*

(b) *How many triangles exist, whose vertices are elements of the set A?*

Solution: (a) Choosing any one specific point from A, we see that there are 12 such segments with this point at one end. For 13 points from A it is therefore possible to form $13 \cdot 12 = 156$ such segments, but every segment was counted twice in this argument, and we therefore only have $156 : 2 = 78$ segments in total. (Of course, this is just a more elementary way to say that we can choose the two endpoints of the line segment at random from the 13 points, yielding a total of $\binom{13}{2} = 78$ line segments.)

(b) Any such a triangle must have two vertices on one line and one on the other. There are $\binom{7}{2} = 21$ ways to choose two points on p, and therefore $21 \cdot 6 = 126$ triangles with two vertices on p. Similarly, there are $\binom{6}{2} = 15$ ways to choose two points on q, and therefore $15 \cdot 7 = 105$ triangles with two vertices on q. This gives us a total of $126 + 105 = 231$ triangles with vertices in A. □

Problem 118. (B-I-2-16) *Determine in how many ways one can assign numbers of the set* $\{1, 2, \ldots, 8\}$ *to the vertices of a cube ABCDEFGH, such that the sum of any two numbers at vertices with a common edge is always an odd number.*

Solution: We first note that vertices of each face of the cube *ABCDEFGH* must be assigned exactly two even and two odd numbers from the given set, with numbers of the same parity always in diagonally opposite vertices in each face. This is true, since we know that if the numbers assigned to vertices on a common edge had the same parity, their sum would be even, contradicting the assumptions of the problem.

If we assign an even number to a vertex A, we will write A_e, and if we assign an odd number to a vertex B, we will write B_o. We can now assume that the vertex A is assigned an even number. The seven vertices of the cube must then necessarily be assigned even or odd numbers from the given set in following way (see picture):

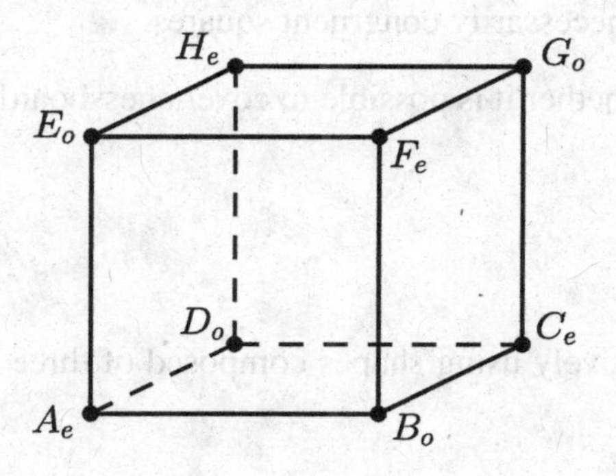

The number of ways to assign the numbers of the set $\{1, 2, \ldots, 8\}$ to the vertices of the cube is then equal to $4! \cdot 4! = (4!)^2 = 576$, since the four even numbers can be assigned in any way, as can the four odd numbers. Similarly, if A is assigned an odd number, we obtain the same number of possibilities.

We see that there exist $2 \cdot (4!)^2 = 1\,152$ assignments of the numbers from the set $\{1, 2, \ldots, 8\}$ to the vertices of the cube $ABCDEFGH$ that satisfy the conditions of the problem. □

4.2. Chessboard Problems

A popular setting for competition problems of all levels of difficulty is the chessboard. Such problems can be set on a standard 8×8 chessboard or a less typical $n \times n$ "chessboard", i.e., a square array consisting of n^2 small squares, or cells. Some of these problems concern possible colourings of the cells, while others might concern themselves with games played under certain rules on the board or ways to cut the square into parts according to some rule. Here are some examples of this type from the Duel.

PROBLEMS

119. Some of the cells of an $n \times n$ chessboard are coloured black. Determine the smallest possible number of cells that must be black such that each 2×2 square on the chessboard contains at least two black cells.

120. Let $n \geq 6$ be an integer. Prove that a given square can be cut into n (not necessarily congruent) squares.

121. Show a manner in which it is possible to cut a given square into (a) 29, (b) 33, (c) 37 not necessarily congruent squares.

122. Determine whether it is possible to cover chessboards with the dimensions

(a) 11×11, and
(b) 12×12

completely, exclusively using shapes composed of three squares as shown in the figure.

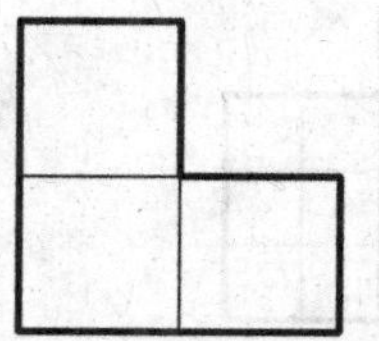

123. Let k be a non-negative integer. For which positive integers $n = 6k+3$ is it possible to cover an $n \times n$ chessboard completely, using only tiles composed of three squares as shown in the figure below?

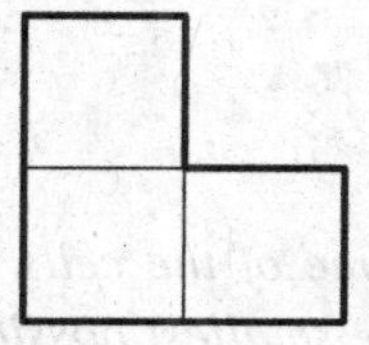

124. We are given a bag full of T-tetrominoes, as shown in the figure below.

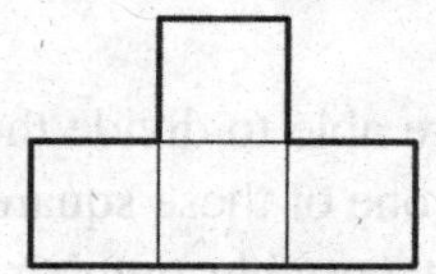

Is it possible to completely cover an 8×12 chessboard with T-tetrominoes without overlapping (assuming that the squares of the chessboard are the same size as the squares of the tetromino)? Is it possible for a 3×8 chessboard? Is it possible for a 7×10 chessboard? In each case, if it is possible, draw such a covering. If not, explain why it is not possible.

125. Determine the number of ways in which a 4×4 chessboard can be covered using eight 2×1 dominos.

126. We are given a board composed of 16 unit squares as shown below. We wish to colour some of the cells green in such a way that, no matter where we place the T-shaped tetromino on the board (with each square of the tetromino covering exactly one square on the board), at least one square of the tetromino will be on a green cell. Determine the smallest possible number of cells we must colour green and prove that this is the smallest number.

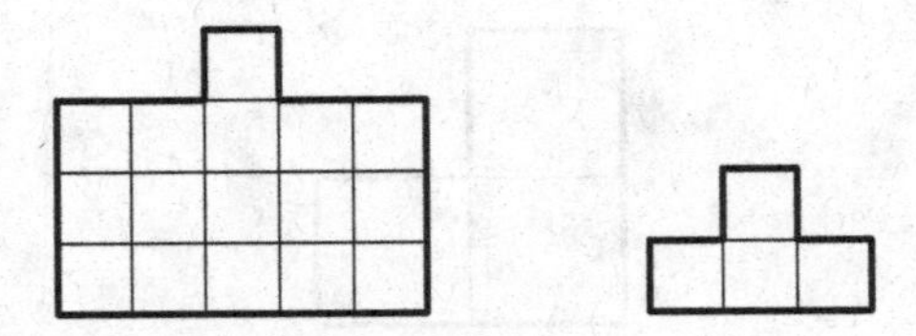

127. We are given a 4×4 array consisting of 16 unit squares. Determine the number of ways in which the array can be covered with five congruent straight triominoes (3×1 rectangles) such that exactly one unit square in the array remains empty.

SOLUTIONS

Problem 119. (A-T-3-09) *Some of the cells of an $n \times n$ chessboard are coloured black. Determine the smallest possible number of cells that must be black, such that each 2×2 square on the chessboard contains at least two black cells.*

Solution: If n is even, we are able to divide the $n \times n$ chessboard into $\frac{n^2}{4}$ squares of size 2×2. Every one of these squares must contain at least two black cells, and the smallest possible number of black cells is therefore at least $\frac{n^2}{2}$. Conventional colouring of the chessboard shows us that such a colouring is indeed possible, and the smallest possible number of black cells is therefore $\frac{n^2}{2}$.

Let us now assume that n is odd. The proof is illustrated below for the case $n = 7$, but the argument works for any odd value of $n \geq 3$.

As in the even case, we draw 2×2 squares on the chessboard, starting from the upper-left and lower-right corners, as shown on the left.

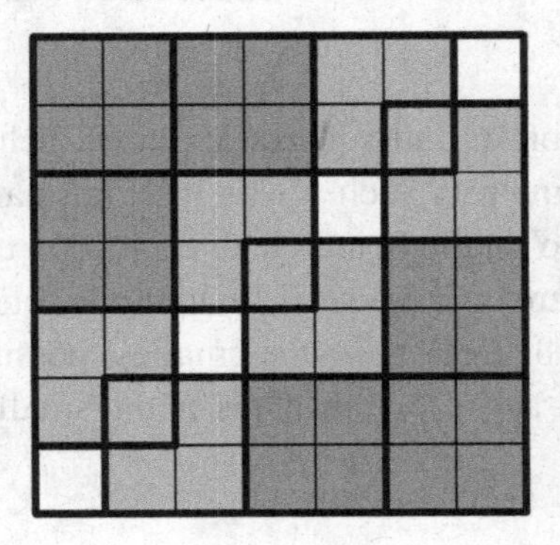

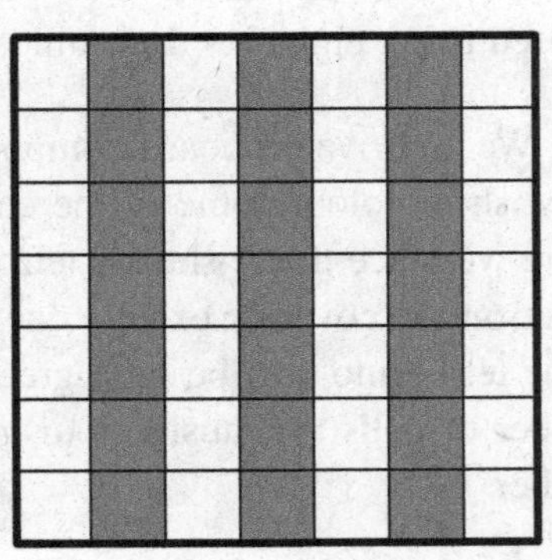

This results in the dark grey squares. The last possible of these overlap in the areas shown in a lighter grey, ordered along the diagonal of the chessboard. It is obvious that at least two cells in each of the dark grey squares must be coloured, and this means we have at least $\frac{n-1}{2} \cdot \frac{n-3}{2} \cdot 2$ coloured squares in this area. For any pair of overlapping light grey squares in the diagonal, we must have at least three coloured cells, i.e., the common cell and one more in each square. This gives us at least another $\frac{n-1}{2} \cdot 3$ squares to colour, and we see that the total number of coloured squares must be at least

$$\frac{n-1}{2} \cdot 3 + \frac{n-1}{2} \cdot \frac{n-3}{2} \cdot 2 = \frac{n(n-1)}{2}.$$

Colouring the chessboard in every other column as shown on the right shows us that this is indeed the smallest possible number of black cells we can colour in order to fulfil the conditions, finishing the proof. □

Problem 120. (I-3-93) *Let $n \geq 6$ be an integer. Prove that a given square can be cut into n (not necessarily congruent) squares.*

Solution: In the following figure, we see how we can cut a square into 6, 7 and 8 squares respectively:

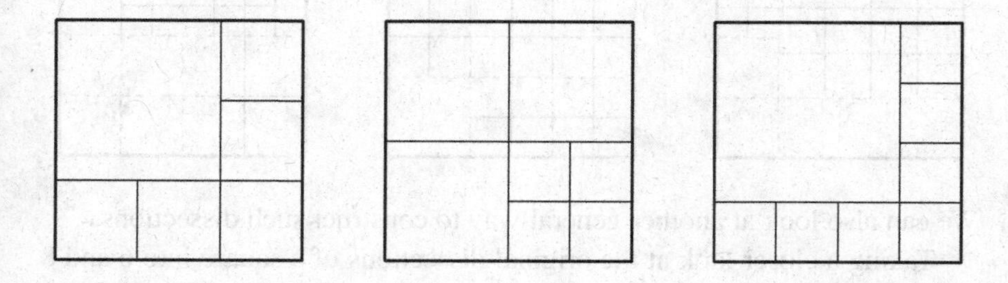

As illustrated in the bottom right of the middle figure (in which the square is cut into 7 smaller squares), we see that the number of squares into which we cut any figure can always be raised by 3, simply by adding the two mid-parallels of the sides. This transforms one square into four, resulting in a net increase of 3 squares. This means that we can certainly cut a square into n squares for any value $n \geq 6$. We simply start with the three given figures, and cutting one square in each of these, results in dissections into 9, 10 and 11 squares respectively; repeating the process results in dissections into 12, 13 and 14 squares, and so on.

Another way to put this is to note that $6 \equiv 0 \pmod 3$, $7 \equiv 1 \pmod 3$ and $8 \equiv 2 \pmod 3$ hold. For any value of $n \geq 6$, we determine its equivalence class modulo 3 and then proceed to add 3 as many times as necessary by dissecting successive squares into 4 smaller squares in each step. □

Problem 121. (C-T-2-94) *Show a manner in which it is possible to cut a given square into* (a) 29, (b) 33, (c) 37 *not necessarily congruent squares.*

Solution: Having just shown that this is certainly possible in the solution to the preceding problem, one way to answer this question is to simply follow the pattern we described there. Noting that $29 \equiv 2 \pmod 3$, $33 \equiv 0 \pmod 3$ and $37 \equiv 1 \pmod 3$ hold, we can start from the divisions of the square into 8, 6 and 7 smaller squares respectively, and create solutions by successive cutting of squares into four equal (smaller) squares. This method is illustrated in the following figure. In order to make the origins visible, the original borders of the 8, 6 and 7 smaller squares have been drawn as dashed lines.

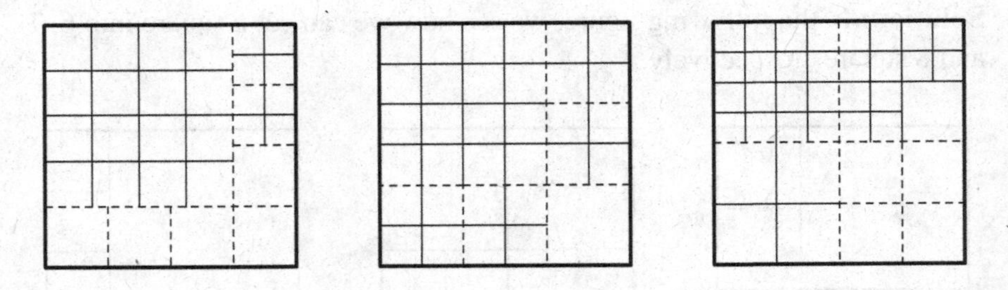

We can also look at another general way to construct such dissections.

Taking a closer look at the original dissections of a square into 6 and 8 small squares, we note that a general method for such dissections can be simply to create a row of equally large small squares on the bottom and right-hand edges of the large square, leaving one large square in the top left.

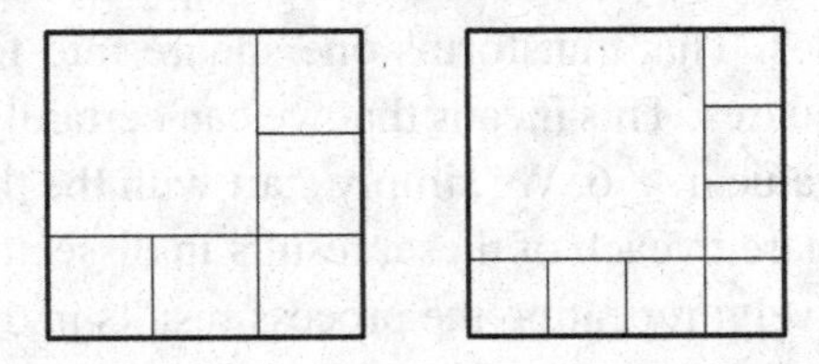

This is possible if we wish to dissect the given square into any even number n of small squares greater than 2, since we can then certainly write $n = 1 + (2k + 1)$ for some positive integer value k. The method then yields $k + 1$ small squares both on the bottom and right-hand side of the original square, all of the same size, along with the larger one remaining in the top left corner. (Note that for $n = 6$ and $n = 8$, we use $k = 2$ and $k = 3$ respectively, writing $6 = 1 + (2 \cdot 2 + 1)$ and $8 = 1 + (2 \cdot 3 + 1)$.)

The number of small squares we are aiming for is, however, odd in each of the three cases of the problem at hand. In order to get around this problem, we can cut the large square in the top left corner into 4 squares. Then, if we want to cut a large square into 29 small squares, for instance, we can write $29 = 4 + (2 \cdot 12 + 1)$, and cut the given square as shown in the left part of the following figure:

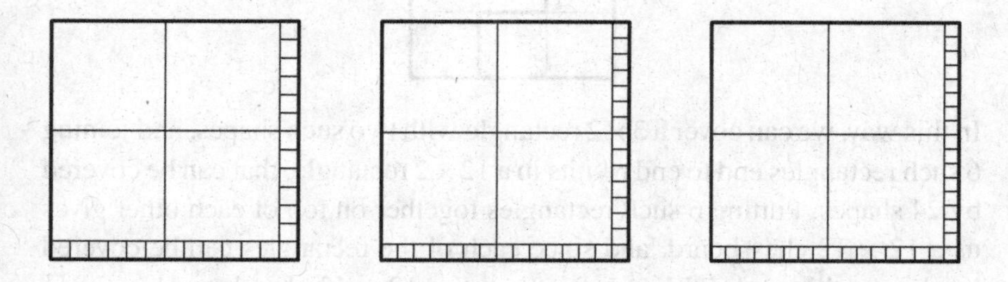

Similarly, writing $33 = 4 + (2 \cdot 14 + 1)$ and $37 = 4 + (2 \cdot 16 + 1)$ gives us the other two dissections in this figure.

Of course, there are many other ways to solve this problem, but these two methods are interesting, as they give some deeper understanding into the general structures of square dissection. □

Problem 122. (C-I-4-98) *Determine whether it is possible to cover chessboards with the dimensions*

(a) 11×11, *and*
(b) 12×12

completely, exclusively using shapes composed of three squares as shown in the figure.

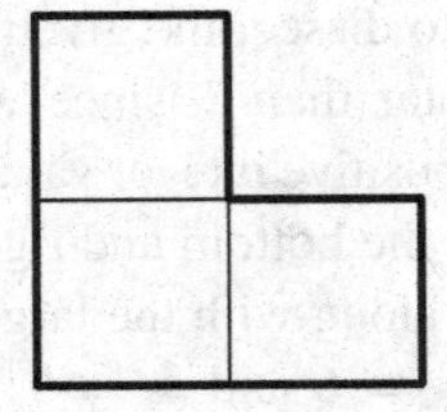

Solution: (a) This is not possible. Each shape covers exactly three squares on the chessboard. Since the number $11 \cdot 11 = 121$ is not divisible by 3, we see that such a covering is certainly impossible.
(b) This is possible; in fact it can be done in many ways. The easiest is achieved by joining two such shapes as shown below.

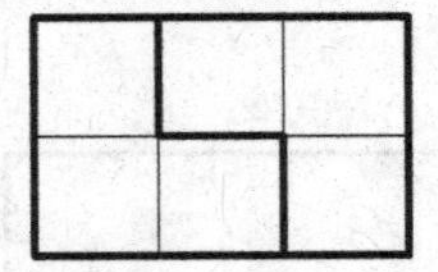

In this way, we can cover a 3×2 rectangle with two such shapes, and joining 6 such rectangles end to end results in a 12×2 rectangle, that can be covered by 24 shapes. Putting 6 such rectangles together on top of each other gives us a 12×12 chessboard, and since each of the rectangles can be covered by shapes, this is also the case for the entire 12×12 chessboard. □

Problem 123. (A-T-3-98) *Let k be a non-negative integer. For which positive integers $n = 6k + 3$ is it possible to cover an $n \times n$ chessboard completely, using only tiles composed of three squares as shown in the figure below?*

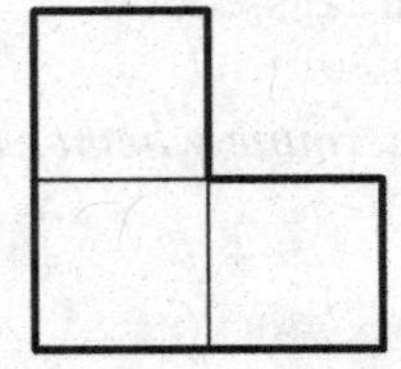

Solution: We first note that this is not possible for $k = 0$. Any tile used in covering a 3×3 chessboard can cover at most one of the four corner squares of the chessboard, and we would therefore require at least four such tiles for a complete covering. However, the total number of squares in this case is only $3 \cdot 3 = 9$, and any possible covering would therefore use three such tiles, yielding a contradiction.

We can now show by induction that it is possible to cover the chessboard in the required way for all integer values $k \geq 1$.

For $k = 1$ (i.e., the 9×9 chessboard), two possible coverings are shown in the figure below. We can therefore assume for the purpose of our induction, that a covering of the $(6k+3) \times (6k+3)$ chessboard is possible for some specific positive integer value of k. We need only to show that this implies a covering for $k+1$, i.e., for the $(6k+9) \times (6k+9)$ chessboard.

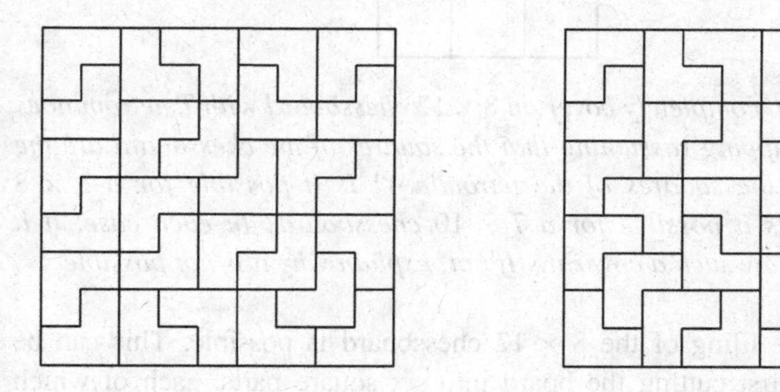

The fact that this is indeed the case is made clear from the following left-hand figure:

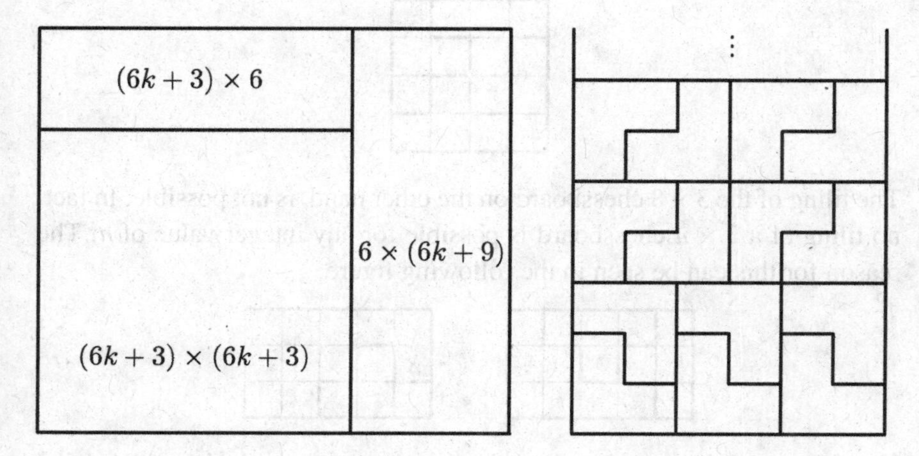

The large $(6k+3) \times (6k+3)$ can be covered by the induction hypothesis. The other two strips are both of width 6 and of odd length of at least 9, and these can certainly be covered as shown on the right. Starting with three 3×2 rectangles, we cover a 3×6 rectangle, and adding as many pairs of 2×3 rectangles as required, allows us to cover any $j \times 6$ rectangle for odd values of $j \geq 3$, completing the proof. □

Problem 124. (C-T-2-02) *We are given a bag full of T-tetrominoes, as shown in the figure below.*

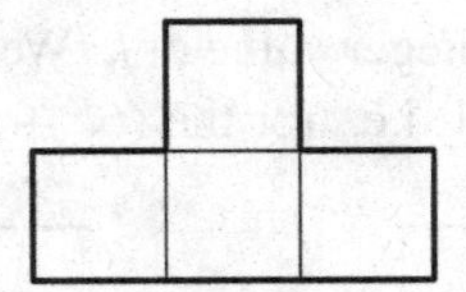

Is it possible to completely cover an 8×12 chessboard with T-tetrominoes without overlapping (assuming that the squares of the chessboard are the same size as the squares of the tetromino)? Is it possible for a 3×8 chessboard? Is it possible for a 7×10 chessboard? In each case, if it is possible, draw such a covering. If not, explain why it is not possible.

Solution: The tiling of the 8×12 chessboard is possible. This can be achieved by first cutting the board into six square parts, each of which measure 4×4 squares. Each of these can be covered by four T-tetrominoes as shown in the following figure:

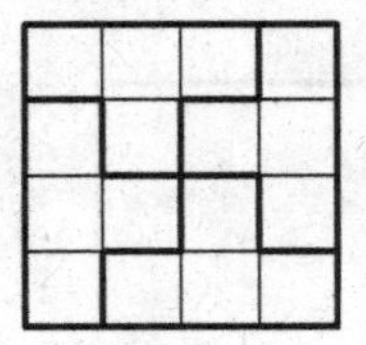

The tiling of the 3×8 chessboard on the other hand, is not possible. In fact, no tiling of a $3 \times n$ chessboard is possible for any integer value of n. The reason for this can be seen in the following figure:

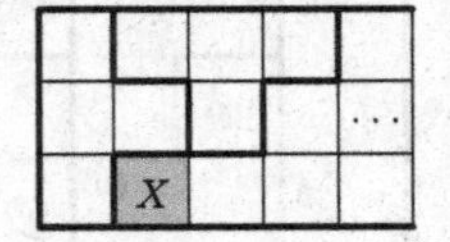

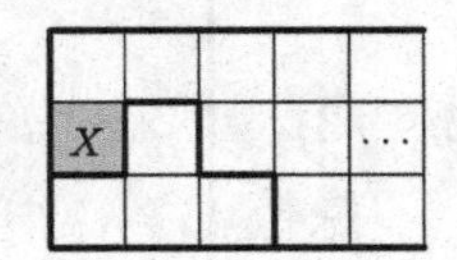

Here, we see the two options for covering the square in the lower left-hand corner of the $3 \times n$ chessboard. In both cases, the grey square marked with an X cannot be covered by a tetromino, making the tiling impossible. (Note that the second tetromino was placed on top in the left figure; the analogous argument holds symmetrically if we place the second tetromino on the bottom.)

Finally, the tiling of the 7×10 chessboard is not possible either. The reason for this lies in the fact that each tile covers four squares, whereas the 7×10 chessboard is composed of $7 \cdot 10 = 70$ squares. Since 70 is not divisible by 4, the tiling cannot be done. □

Problem 125. (C-T-1-07) *Determine the number of ways in which a* 4×4 *chessboard can be covered using eight* 2×1 *dominos.*

Solution: The total number of such coverings equals 36. There are many ways to show this, but one way is the following.

We consider two cases. If the chessboard can be divided into two equal pieces by a vertical line in the middle of the board as shown in the figure below, each half can be covered independently by four dominos.

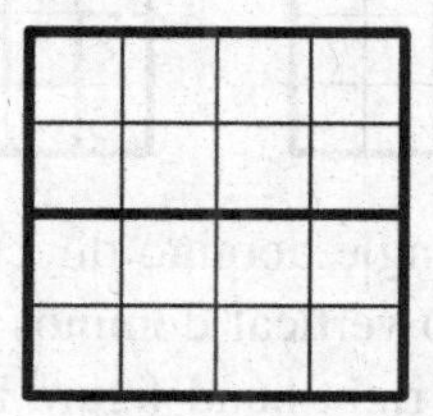

There are five ways to cover a 2×4 rectangle with four dominos, and these are illustrated below.

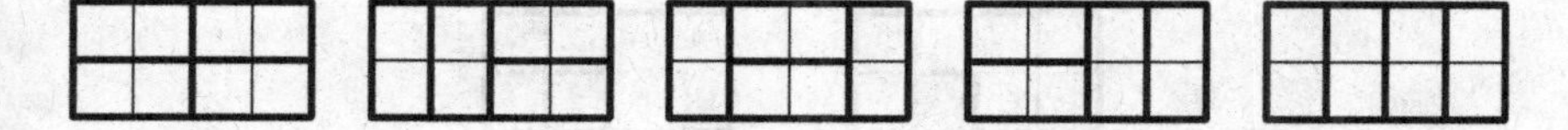

Since any combinations of these five possibilities are possible, this first case yields $5 \cdot 5 = 25$ different coverings.

We now consider the case in which the central division is not possible. In this case, there must be a domino that straddles the centre horizontal line, and one such domino must be the left-most. If this domino is on the left edge, we have two possibilities, as shown in the following figure.

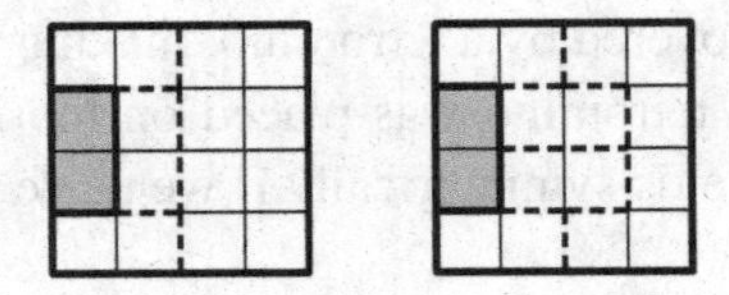

The left-hand corner squares can only be covered in one way, and the necessary dominos are shown with dashed borders. On the left, we see the option in which a second domino is placed vertically next to the one on the left. In this case, there exists a vertical line dividing the 4×4 square in half, and just as was shown in the first case, there are five ways to cover the right-hand 4×2 rectangle. On the right, we see the second option, in which two dominos are placed horizontally next to the one on the left. In this case, there is only one way to finish the covering with three more dominos. Altogether, this option yields six coverings.

The next option is for the left-most vertical domino to be in the second column from the left, as shown in the following figure.

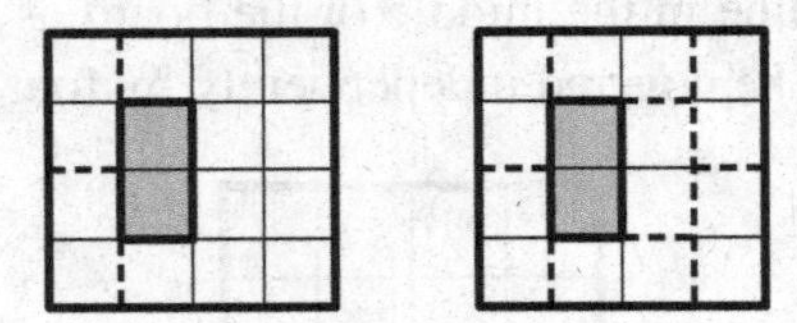

Since there may not be a single domino directly to the left of this one in this case, there must be two vertical dominos placed to its left, as shown in the left-hand figure. The right-hand figure is therefore the only way to complete the covering in this case.

Next, the left-most vertical domino can be placed in the third column from the left, as shown below.

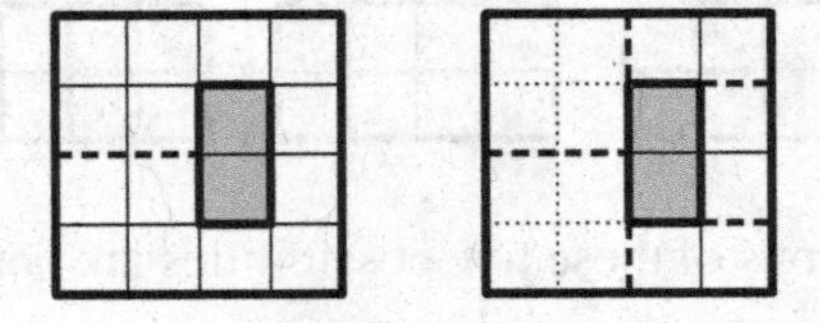

Since we have assumed that the two left-most columns do not contain a vertical domino covering the central horizontal line (a fact that is symbolised in the left figure by the dashed line), we see that any possible covering in this case must include the four dominos as shown on the right. The left half of the 4×4 chessboard is composed of two 2×2 boxes, and we can cover each of these with two dominos either vertically or horizontally (the possible divisions between them are shown as dotted lines), yielding a total of $2 \cdot 2 = 4$ possible coverings in this case.

Finally, as suggested by the following figure, there is no possible covering in which the left-most vertical domino is on the right edge.

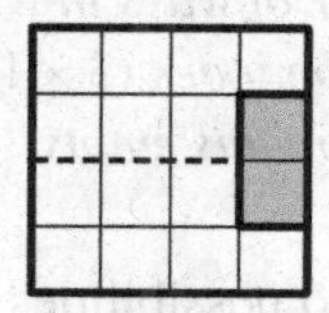

The horizontal dividing line creates two sections with 7 squares each, and since 7 is odd, these sections cannot be covered by dominos.
Summing up all the possibilities, we see that there are indeed $25 + 6 + 1 + 4 = 36$ coverings, as claimed. □

Problem 126. (C-T-3-08) *We are given a board composed of* 16 *unit squares as shown below. We wish to colour some of the cells green in such way that, no matter where we place the* T*-shaped tetromino on the board (with each square of the tetromino covering exactly one square on the board), at least one square of the tetromino will be on a green cell. Determine the smallest possible number of cells we must colour green and prove that this is the smallest number.*

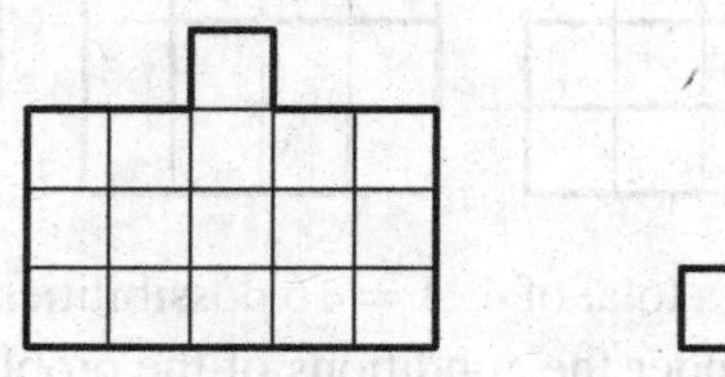

Solution: The smallest number of cells is 4. If we colour the four cells as shown in the diagram, any placement of the polyomino will cover one of these cells.

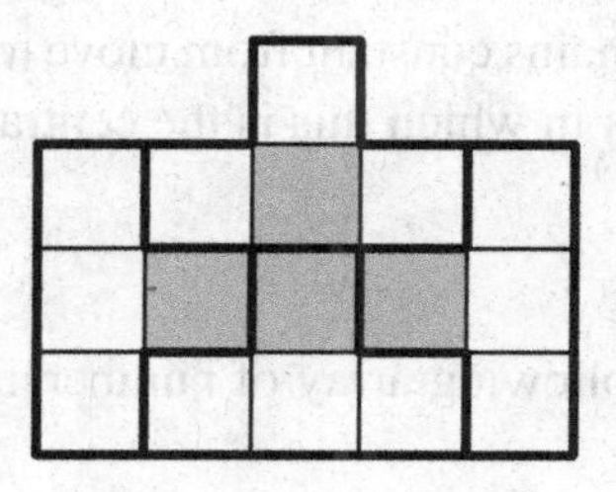

No smaller number is possible, since the board can be cut into the four parts shown in the shape of the given polyomino, no two of which have a common cell. □

Problem 127. (C-T-1-14) *We are given a* 4×4 *array consisting of* 16 *unit squares. Determine the number of ways in which the array can be covered with five congruent straight triominoes* (3×1 *rectangles*), *such that exactly one unit square in the array remains empty.*

Solution: We can consider two possibilities for the position of the empty cell by covering the given 4×4 table by five straight triominoes. This empty square can either be in a corner of the 4×4 table or not. It is quite easy to check that there is no possible way to cover if the empty square is either a central square or a central edge square. This means that the empty square must be in a corner.

Since the square table is symmetric with respect to its centre, we can assume without loss of generality that the empty cell is left over at the bottom left corner. In this case, we get the four possibilities shown in the picture.

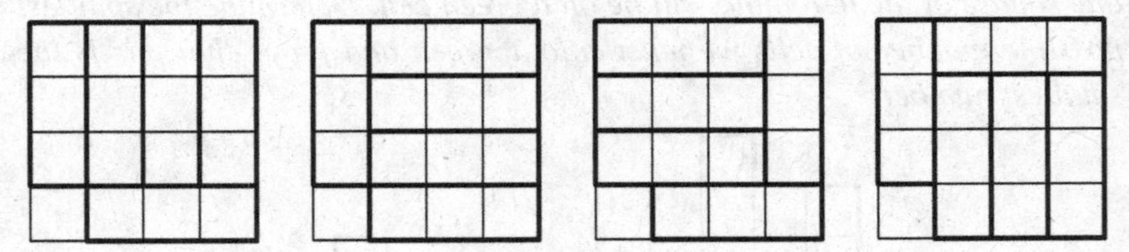

This means that there exist a total of $4 \cdot 4 = 16$ possibilities for the covering of the given square table under the conditions of the problem. □

4.3. Invariants

Many problems having to do with mathematical games require the solver to find some aspect that remains constant from move to move, i.e., an invariant. Here are a few problems in which this is the central theme.

PROBLEMS

128. We are given the following array of numbers:

1	2	3
2	3	1
3	1	2

In each move we are allowed to add 1 to all three numbers in any row or to subtract 1 from each of the three numbers in any column. Which of the following arrays can be the result of a number of such moves? If it is possible to reach this array, determine the moves that must be made. If it is not possible, explain why not.

2	2	5
1	1	1
4	2	2

2	2	3
2	2	1
4	1	3

129. One number from the set $\{-1, 1\}$ is written in each vertex of a square. In one step, the number in each vertex is replaced by the product of this number and the two numbers in adjacent vertices. Prove that it is possible to reach a situation in which the number 1 is written in each vertex after a certain number of steps if and only if the number 1 is already written in all four vertices from the start.

130. Red, green and blue fireflies live on the magic meadow.

- If two blue fireflies meet, they change into one red firefly.
- If three red fireflies meet, they change into one blue firefly.
- If three green fireflies meet, they change into one red firefly.
- If one red and one green firefly meet, they change into one blue firefly.
- If one red and one blue firefly meet, they change into two green fireflies.

Initially, there are 2014 red fireflies on the meadow. At some point, there are exactly five fireflies on the meadow. Prove that these five fireflies do not all have the same colour. Is it possible for exactly one firefly to remain on the meadow?

131. A cube F with edge length 9 is divided into 9^3 small cubes with edges of the length 1 by planes parallel to its faces. One small cube is removed from the centre of each of the six faces of the cube F. Let us denote the resulting solid by G. Is it possible to build the solid G only from rectangular cuboids with the dimensions $1 \times 1 \times 3$?

SOLUTIONS

Problem 128. (C-I-4-02) *We are given the following array of numbers:*

1	2	3
2	3	1
3	1	2

In each move we are allowed to add 1 *to all three numbers in any row or to subtract* 1 *from each of the three numbers in any column. Which of the following arrays can be the result of a number of such moves? If it is possible to reach this array, determine the moves that must be made. If it is not possible, explain why not.*

2	2	5
1	1	1
4	2	2

2	2	3
2	2	1
4	1	3

Solution: It is not possible to reach either of these two arrays. In order to see why, we consider the sum of all numbers in each of the three arrays. The sum of all numbers in the given starting array is 18, while the sum of all numbers in each of the two arrays we wish to reach is 20. We note that either of the two allowed moves changes the sum of the numbers in the array by 3. Since the starting sum of 18 is divisible by 3, this means that any sum that is not divisible by 3 can never be reached with the allowed moves. Since 20 is not divisible by 3, we see that neither array can be reached, as claimed.

□

Problem 129. (B-I-4-14) *One number from the set* $\{-1, 1\}$ *is written in each vertex of a square. In one step, the number in each vertex is replaced by the product of this number and the two numbers in adjacent vertices. Prove that it is possible to reach a situation in which the number* 1 *is written in each vertex after a certain number of steps if and only if the number* 1 *is already written in all four vertices from the start.*

Solution: Let us consider a square $A_1A_2A_3A_4$. Let a_i be numbers of the set $\{-1; 1\}$ written at the vertices A_i $(i = 1, 2, 3, 4)$ in some step. Let us define $k = a_1a_2a_3a_4$. In the following step, we will then have the numbers

ka_{i+2} at the vertices A_i (with $a_5 = a_1$, $a_6 = a_2$). Moreover, we see that the product $k^5 = k$ of all four numbers remains the same from step to step.

Therefore, if we get numbers 1 at all vertices in some step, then it must be $a_1 = a_2 = a_3 = a_4$ in the previous step. Therefore we get $k = 1$, which follows $a_1 = a_2 = a_3 = a_4 = 1$. Thus, by this way we get numbers 1 at all vertices initially. On the other hand in this case we can see that the numbers at all vertices are 1 in all steps. □

Problem 130. (A-T-2-14) *Red, green and blue fireflies live on the magic meadow.*

- *If two blue fireflies meet, they change into one red firefly.*
- *If three red fireflies meet, they change into one blue firefly.*
- *If three green fireflies meet, they change into one red firefly.*
- *If one red and one green firefly meet, they change into one blue firefly.*
- *If one red and one blue firefly meet, they change into two green fireflies.*

Initially, there are 2014 *red fireflies on the meadow. At some point, there are exactly five fireflies on the meadow. Prove that these five fireflies do not all have the same colour. Is it possible for exactly one firefly to remain on the meadow?*

Solution: Let us consider the meadow with r red, g green and b blue fireflies. Let I be the remainder of the term $r+2g+3b$ after division by the number 5. By direct verification we can show that I is invariant. Initially, we have $r = 2014$, $b = g = 0$, and thus $I = 4$.

If five fireflies of the same colour remain on the meadow, we have $I = 0$, which contradicts $I = 4$.

If one firefly remains on the meadow, we have $I = 1$ for the red firefly, $I = 2$ for the green one and $I = 3$ for the blue firefly. All cases contradict $I = 4$. So only one firefly cannot remain on the meadow. □

Problem 131. (A-T-2-16) *A cube F with edge length* 9 *is divided into* 9^3 *small cubes with edges of the length* 1 *by planes parallel to its faces. One small cube is removed from the centre of each of the six faces of the cube F. Let us denote the resulting solid by G. Is it possible to build the solid G only from rectangular cuboids with the dimensions* $1 \times 1 \times 3$?

Solution: Let us place our cube F in a system of coordinates as shown in the figure below. Each of the small cubes can be uniquely determined by three coordinates $[x, y, z]$ in a similar manner as points. The cube with a vertex at $(0, 0, 0)$ thus gets coordinates $[0, 0, 0]$ and the cube with a vertex $(9, 9, 9)$ gets coordinates $[8, 8, 8]$. Let us colour each of the small cubes with one of three colours depending of the remainder obtained after division of the sum of its coordinates $x + y + z$ by 3.

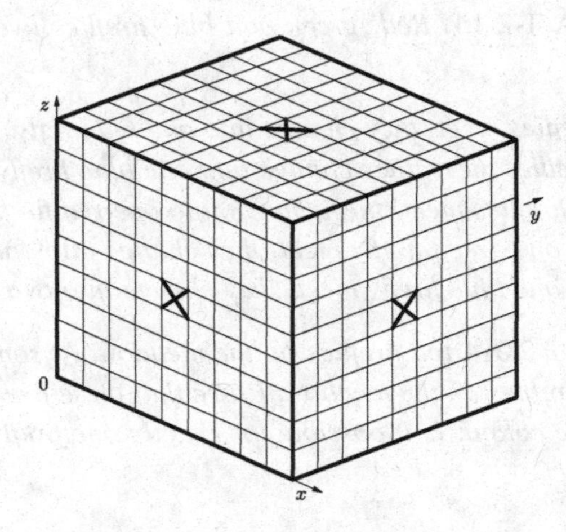

The cube F then consists of the same number of small cubes of each colour. The cubes removed from F (three of these are marked with crosses in the figure) have coordinates $[4, 4, 0]$, $[4, 4, 8]$, $[0, 4, 4]$, $[8, 4, 4]$, $[4, 0, 4]$ and $[4, 8, 4]$, and the remainders obtained after dividing these sums by 3 are 2, 1, 2, 1, 2 and 1, respectively. We see that there are more small cubes in the solid G for which the corresponding remainder is 0 than there are with remainder 1 and 2.

On the other hand each of the rectangular cuboids can only have three small cubes of three distinct colours. It follows that it is not possible to build G from such cuboids, as was claimed. □

4.4. Properties of Points

The last problems we take a look at in this section all concern certain properties of points. The points in question can be spread around haphazardly in the plane or in three-dimensional space or lie in a regular grid of some kind, and the properties in question can concern such things as possible colourings or distances. A tool that can often be applied to such problems is the pigeon-hole principle, and we shall encounter some examples of this.

PROBLEMS

132. Every point of the plane is coloured in one of two colours. Prove that there exist among them two points of the same colour, such that their distance is equal to 1996.

133. Each vertex of a regular hexagon can be coloured either white or red. In how many different ways can the vertices of the hexagon be painted? (We consider two paintings as different if and only if there is no isometric mapping from one painting to the other.)

134. Each vertex of a regular hexagon $ABCDEF$ is coloured in one of three colours (red, white or blue), such that each colour is used exactly twice. Determine in how many ways we can do this, if any two adjacent vertices of the hexagon are always coloured in distinct colours.

135. Some of the vertices of a cube are coloured in such a way that no four coloured vertices lie in a common plane. Determine the largest possible number of coloured vertices.

136. We are given 385 different points in the interior of a $3\,\text{m} \times 2\,\text{m}$ rectangle. Prove that there exist five points among these which can be covered by a square with sides of length 25 cm.

137. The floor of a room with area 5 m^2 is (partly or completely) covered using nine carpets, each of which has an area of exactly 1 m^2, but the shapes of which are arbitrary. Prove that there must exist two carpets that cover a common section of floor with an area of at least $\frac{1}{9}$ m^2.

138. We are given 1300 points inside a unit sphere. Prove that there exists a sphere with radius $\frac{2}{9}$ containing at least 4 of these points.

SOLUTIONS

Problem 132. (B-I-2-96) *Every point of the plane is coloured in one of two colours. Prove that there exist among them two points of the same colour, such that their distance is equal to* 1996.

Solution: This is a very elementary application of the pigeon-hole principle. If we consider the vertices of any equilateral triangle with sides of length 1996, at least two of them must be of the same colour, proving the claim. □

Problem 133. (B-T-1-98) *Each vertex of a regular hexagon can be coloured either white or red. In how many different ways can the vertices of the hexagon be painted*? (*We consider two paintings as different if and only if there is no isometric mapping from one painting to the other.*)

Solution: We can consider the following cases:

Case I: 0 or 1 vertex is white. In each case, there is one possible colouring.

Case II: 1 or 5 vertices are white. In each case, there is once again only one possible colouring.

Case III: 2 or 4 vertices are white. If 2 vertices are white, there can be 0, 1 or 2 red vertices between them or the shorter side. The same holds if there are 4 white vertices for the two red ones. There are therefore 3 colourings in each of these cases.

Case IV: 3 vertices are white and 3 are red. In this case, all three white vertices can be in a row (i.e. with no red vertices in between), or two can be joined by a common edge and one separated on one side by a red vertex (and thus by two red vertices on the other side), or white and red vertices can alternate. This yields a total of three possible colourings in this case.

Altogether, this gives us $2 \cdot 1 + 2 \cdot 1 + 2 \cdot 3 + 3 = 13$ colourings. □

Problem 134. (C-I-2-16) *Each vertex of a regular hexagon $ABCDEF$ is coloured in one of three colours* (*red, white or blue*), *such that each colour is used exactly twice. Determine in how many ways we can do this*, *if any two adjacent vertices of the hexagon are always coloured in distinct colours.*

Solution: Let us denote the three colours by capitals, R (red), W (white) and B (blue). Let us first assume that the vertex A of the considered hexagon

ABCDEF is coloured red (**R**). All possibilities fulfilling the conditions of the given problem are listed in the table below (the vertex A is repeated in the right column).

A	B	C	D	E	F	A
R	W	**R**	B	W	B	**R**
R	B	**R**	W	B	W	**R**
R	W	B	**R**	W	B	**R**
R	W	B	**R**	B	W	**R**
R	B	W	**R**	W	B	**R**
R	B	W	**R**	B	W	**R**
R	B	W	B	**R**	W	**R**
R	W	B	W	**R**	B	**R**

We see that we obtain a total of eight possibilities if vertex A is coloured red. Similarly, if we colour A either white or blue, we obtain another eight possibilities in each case. In total, there therefore exist $3 \times 8 = 24$ possible colourings of all vertices of the hexagon *ABCDEF*. □

Problem 135. (C-I-3-99) *Some of the vertices of a cube are coloured in such a way that no four coloured vertices lie in a common plane. Determine the largest possible number of coloured vertices.*

Solution: We shall prove that five is the largest possible number of vertices that can be coloured.

We consider a cube with the vertices numbered as shown in the figure below.

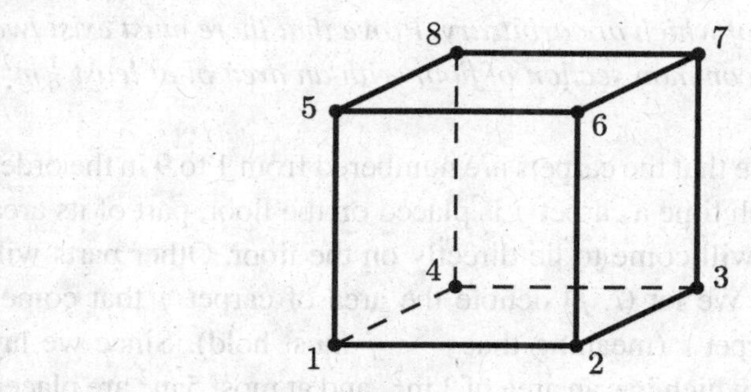

We assume that at least five vertices of the cube are coloured. In this case, at least three must lie in one of two parallel faces of the cube, and without loss of generality, we can assume that these are in the bottom plane. Also without loss of generality, we can assume that the coloured points are 1, 2 and 3. (Otherwise we can rotate and possibly reflect the cube accordingly.) We now note that 4 cannot be coloured due to the constraints of the problem. We now see that the top surface cannot have two consecutive colored vertices. This is seen by considering all possible pairs:

- 5 and 6: 1, 2, 5, 6 lie in a common plane;
- 6 and 7: 2, 3, 6, 7 lie in a common plane;
- 7 and 8: 1, 2, 7, 8 lie in a common plane;
- 5 and 8: 2, 3, 5, 8 lie in a common plane.

It therefore follows that six vertices of the cube cannot be coloured. On the other hand, if 6 and 8 are coloured along with 1, 2 and 3, no four of these five vertices lie in a common plane. We see that 5 is indeed the largest possible number of coloured vertices, as claimed. □

Problem 136. (C-I-2-03) *We are given* 385 *different points in the interior of a* $3\,m \times 2\,m$ *rectangle. Prove that there exist five points among these which can be covered by a square with sides of length* 25 *cm.*

Solution: Since 25 cm equals $\frac{1}{4}$ m, we can divide the rectangle into $12 \cdot 8 = 96$ squares with sides of length 25 cm. Noting that $385 = 4 \cdot 96 + 1$, the pigeon-hole principle tells us that there must exist one such square covering at least 5 of the given points, as claimed. □

Problem 137. (B-I-4-02) *The floor of a room with area* $5\,m^2$ *is (partly or completely) covered using nine carpets, each of which has an area of exactly* $1\,m^2$, *but the shapes of which are arbitrary. Prove that there must exist two carpets that cover a common section of floor with an area of at least* $\frac{1}{9}\,m^2$.

Solution: We imagine that the carpets are numbered from 1 to 9 in the order they are laid out. Each time a carpet j is placed on the floor, part of its area (possibly of area 0) will come to lie directly on the floor. Other parts will lie on other carpets. We let (i, j) denote the area of carpet j that comes to lie directly on carpet i (meaning that $i < j$ must hold). Since we lay nine carpets, each of which has an area of $1\,\mathrm{m}^2$, and at most $5\,\mathrm{m}^2$ are placed directly on the floor, it follows that the sum of all areas (i, j) must be greater than or equal to $4\,\mathrm{m}^2$. Since there are nine carpets, there are $\binom{9}{2} = 36$ such areas, and it follows from the pigeon-hole principle that at least one of these must have an area of at least $\frac{4}{36} = \frac{1}{9}$, as claimed. □

Another solution: We can prove the claim by contradiction. We therefore assume that no two carpets cover a section with an area of $\frac{1}{9}$ m^2. If we place the first carpet on the floor, it covers an area of 1 m^2. Placing the second carpet on the floor, it must then cover an area of more than $\frac{8}{9}$ m^2 of the floor, which is not already covered. Placing the third carpet on the floor, this must then cover an area of more than $\frac{7}{9}$ m^2 of the floor, which is not already covered, and so on. Finally, placing the ninth carpet on the floor, it must cover an area of more than $\frac{1}{9}$ m^2 of the floor, which is not already covered. The nine carpets together then cover a total area larger than

$$1 + \frac{8}{9} + \frac{7}{9} + \frac{6}{9} + \frac{5}{9} + \frac{4}{9} + \frac{3}{9} + \frac{2}{9} + \frac{1}{9} = 5\text{m}^2.$$

This contradicts the fact that the area of the room is only 5 m^2, and there must therefore exist two carpets covering a common section with an area of at least $\frac{1}{9}$ m^2, as claimed. □

Problem 138. (A-T-1-00) *We are given* 1300 *points inside a unit sphere. Prove that there exists a sphere with radius* $\frac{2}{9}$ *containing at least* 4 *of these points.*

Solution: We first note that the unit sphere can be circumscribed by a cube with edges of length 2. This cube can be cut into $8^3 = 512$ small cubes with edges of length $\frac{1}{4}$. As we see in the following figure, the small cubes sharing an outer edge with the larger circumscribed cube all lie completely outside the original unit sphere.

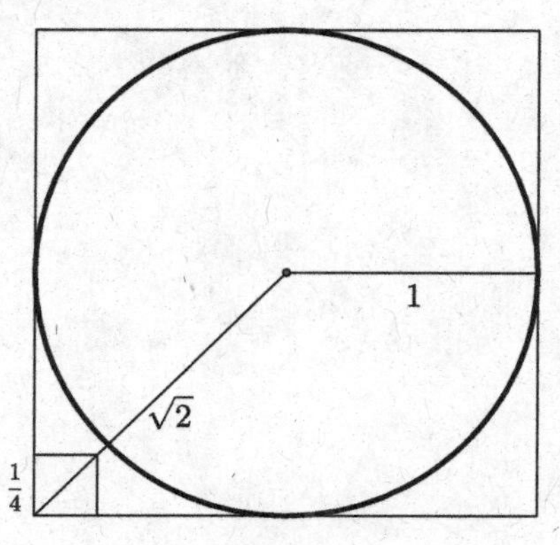

This is true because

$$\sqrt{2} - 1 > \tfrac{1}{4}\sqrt{2} \quad \Leftrightarrow \quad \tfrac{3}{4}\sqrt{2} > 1 \quad \Leftrightarrow \quad \tfrac{9}{16} \cdot 2 > 1$$

certainly holds. Eliminating these $12 \cdot 6 + 8 = 80$ small edge cubes, we see that the remaining $512 - 80 = 432$ small cubes cover the sphere completely. Each of these small cubes has a circumradius of $\frac{1}{8}\sqrt{3}$, and since

$$\tfrac{1}{8}\sqrt{3} < \tfrac{2}{9} \quad \Leftrightarrow \quad \tfrac{3}{64} < \tfrac{4}{81} \quad \Leftrightarrow \quad 243 < 256$$

certainly holds, the 432 spheres with midpoints in the centres of the small cubes and radii $\frac{2}{9}$ completely cover the given unit sphere.

In order to complete the proof, it now only remains to note that $1300 = 3 \cdot 432 + 4$, and it follows from the pigeon-hole principle that at least one of these small spheres contains at least four of the given points, as claimed. □

Chapter 5

Geometry

5.1. Geometric Inequalities

Nestled somewhere between the classic areas of geometry and inequalities, we find the quite popular topic of geometric inequalities. Problems in this area can pertain to relationships between the lengths of certain line segments in a particular configuration, areas of certain specific regions, or much more. Here are some interesting examples.

PROBLEMS

139. Let u and v be the distances of an arbitrary point on the side AB of an acute-angled triangle ABC to its sides AC and BC. Furthermore, let h_a and h_b be the lengths of the altitudes from its vertices A and B, respectively. Prove that the following inequalities hold:

$$\min\{h_a, h_b\} \leq u + v \leq \max\{h_a, h_b\}.$$

140. Prove that the sum of the lengths of two segments connecting the midpoints of opposite sides of an arbitrary quadrilateral is less than the sum of the lengths of its diagonals.

141. Let us consider an acute-angled triangle ABC. Let D, E, F be the feet of the altitudes from the vertices A, B, C, respectively. Furthermore, let K, L, M denote the points of intersection of the lines AD, BE, CF with the circumcircle of the triangle ABC different from the vertices A, B, C, respectively. Prove that the following inequality holds:

$$\min\left\{\frac{|KD|}{|AD|}, \frac{|LE|}{|BE|}, \frac{|MF|}{|CF|}\right\} \leq \frac{1}{3}.$$

142. We are given a common external tangent t to circles $c_1(O_1, r_1)$ and $c_2(O_2, r_2)$, which have no common point and lie in the same half-plane defined by t. Let d be the distance between the tangent points of circles c_1 and c_2 with t. Determine the smallest possible length of a broken line AXB (i.e., the union of line segments AX and XB), such that A lies on c_1, B lies on c_2, and X lies on t.

143. Let us consider the triangle ABC with altitudes $h_a = 24$ and $h_b = 32$. Prove that its third altitude h_c fulfils the inequalities

$$13 < h_c < 96.$$

144. Let $2s$ be the length of the perimeter of a triangle ABC and let ρ, r_a, r_b and r_c be the radii of the incircle and three excircles of ABC respectively. Prove that the following inequality holds:

$$\sqrt{\rho \cdot r_a} + \sqrt{\rho \cdot r_b} + \sqrt{\rho \cdot r_c} \leq s.$$

145. We consider a line segment AB of the length c and all right triangles with hypotenuse AB. For all such right triangles, determine the maximum diameter of a circle with the centre on AB which is tangent to the other two sides of the triangle.

146. Let a, b, c be the lengths of the sides of a triangle. Prove that the following inequality holds:

$$3a^2 + 2bc > 2ab + 2ac.$$

SOLUTIONS

Problem 139. (A-I-3-15) *Let u and v be the distances of an arbitrary point on the side AB of an acute-angled triangle ABC to its sides AC and BC. Furthermore, let h_a and h_b be the lengths of the altitudes from its vertices A and B, respectively. Prove that the following inequalities hold:*

$$\min\{h_a, h_b\} \leq u + v \leq \max\{h_a, h_b\}.$$

Solution: Let P be an arbitrary interior point on the segment AB. Let point $G \in BC$ such that $PG \perp BC$ and point $D \in AC$ such that $PD \perp AC$. Then $u = |PD|$ and $v = |PG|$.

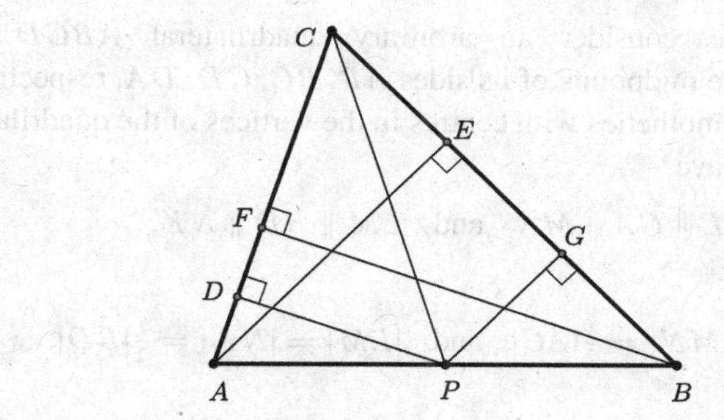

Now let AE and BF be altitudes of the triangle ABC. Then

$$2S_{ABC} = |BC| \cdot |AE| = |AC| \cdot |BF|,$$

where S_{ABC} denotes the area of the triangle ABC. Dividing ABC into two triangles APC and PBC, we obtain

$$2S_{ABC} = |AC| \cdot |PD| + |BC| \cdot |PG|,$$

and therefore

$$|BC| \cdot |AE| = |AC| \cdot |PD| + |BC| \cdot |PG| = |AC| \cdot |BF|.$$

Without loss of generality, we can assume $|AC| \leq |BC|$. By the above equality we have

$$|BC| \cdot |AE| = |AC| \cdot |PD| + |BC| \cdot |PG| \leq |BC| \cdot |PD| + |BC| \cdot |PG|,$$

and therefore $|AE| \leq |PD| + |PG|$. Furthermore, the equality also yields

$$|AC| \cdot |BF| = |AC| \cdot |PD| + |BC| \cdot |PG| \geq |AC| \cdot |PD| + |AC| \cdot |PG|,$$

which gives us $|BF| \geq |PD| + |PG|$, and therefore

$$|AE| \leq |PD| + |PG| \leq |BF|.$$

Because of $|AC| \leq |BC|$, the altitudes AE and FB fulfil the inequality $|FB| \geq |AE|$. We therefore have

$$\min\{h_a, h_b\} = |AE| \quad \text{and} \quad \max\{h_a, h_b\} = |FB|$$

and the inequalities hold, completing the proof. □

Problem 140. (B-I-3-14) *Prove that the sum of the lengths of two segments connecting the midpoints of opposite sides of an arbitrary quadrilateral is less than the sum of the lengths of its diagonals.*

Solution: Let us consider an arbitrary quadrilateral $ABCD$. Let K, L, M, N be the midpoints of its sides AB, BC, CD, DA, respectively. Because of the homotheties with centres in the vertices of the quadrilateral with ratio $\frac{1}{2}$, we have

$$KL \parallel CA \parallel MN \quad \text{and} \quad LM \parallel BD \parallel NK,$$

and

$$|KL| = |MN| = \tfrac{1}{2}|AC| \quad \text{and} \quad |LM| = |NK| = \tfrac{1}{2}|BD|. \qquad (1)$$

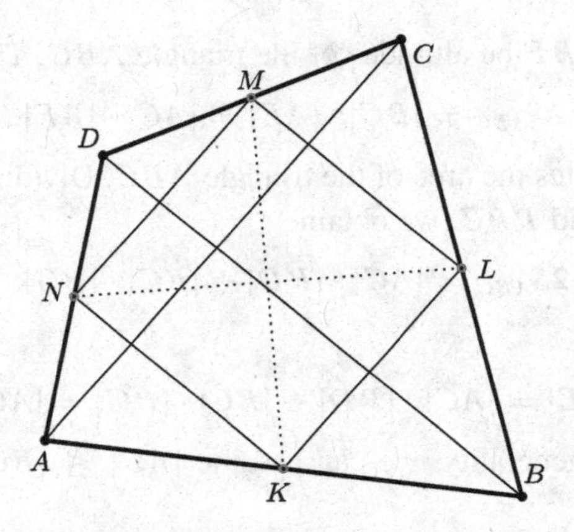

This implies that $KLMN$ is a parallelogram (the *Varignon* parallelogram). Applying the triangle inequality to its diagonals, we obtain

$$|KM| < |KL| + |LM| \quad \text{and} \quad |LN| < |LM| + |MN|.$$

Summing up last two inequalities and applying (1) then yields

$$|KM| + |LN| < (|KL| + |MN|) + 2|LM| = |AC| + |BD|,$$

which completes the proof. □

Problem 141. (A-T-2-12) *Let us consider an acute-angled triangle ABC. Let D, E, F be the feet of the altitudes from the vertices A, B, C, respectively. Furthermore, let K, L, M denote the points of intersection of the lines AD, BE, CF with the circumcircle of the triangle ABC different from the vertices A, B, C, respectively. Prove that the following inequality holds:*

$$\min\left\{\frac{|KD|}{|AD|}, \frac{|LE|}{|BE|}, \frac{|MF|}{|CF|}\right\} \leq \frac{1}{3}.$$

Solution: Let V be the orthocentre of the acute-angled triangle ABC. It is well known that the mirror images of V with respect to the lines AB, BC and CA lie on the circumcircle of this triangle. These are the points M, K and L, respectively. Letting S_{BCV} and S_{ABC} denote the areas of the triangles BCV and ABC respectively, we therefore have

$$\frac{S_{BCV}}{S_{ABC}} = \frac{|VD|}{|AD|} = \frac{|KD|}{|AD|}.$$

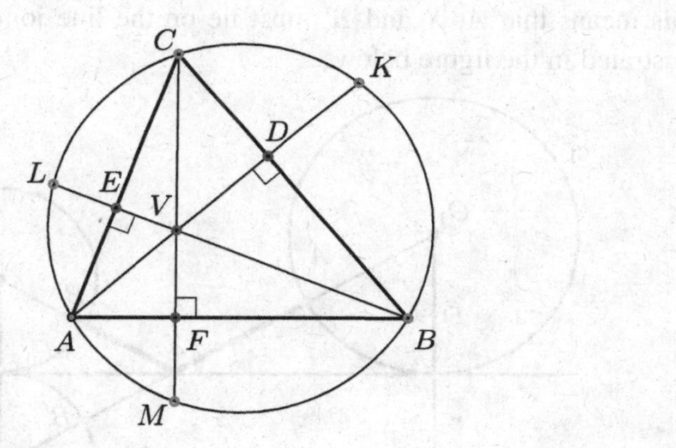

Similarly, we also obtain

$$\frac{S_{CAV}}{S_{ABC}} = \frac{|VE|}{|BE|} = \frac{|LE|}{|BE|} \quad \text{and} \quad \frac{S_{ABV}}{S_{ABC}} = \frac{|VF|}{|CF|} = \frac{|MF|}{|CF|},$$

and adding the left-hand sides of these last three equalities gives us

$$\frac{S_{ABV}}{S_{ABC}} + \frac{S_{CAV}}{S_{ABC}} + \frac{S_{BCV}}{S_{ABC}} = \frac{S_{ABV} + S_{CAV} + S_{BCV}}{S_{ABC}} = 1.$$

Summing the right-hand sides therefore gives us

$$\frac{|KD|}{|AD|} + \frac{|LE|}{|BE|} + \frac{|MF|}{|CF|} = 1.$$

Since each of the three fractions in this sum is a positive real number, not all three values can be greater than $\frac{1}{3}$, and the proof is complete. □

Problem 142. (B-I-4-11) *We are given a common external tangent t to circles $c_1(O_1, r_1)$ and $c_2(O_2, r_2)$, which have no common point and lie in the same half-plane defined by t. Let d be the distance between the tangent* *points of circles c_1 and c_2 with t. Determine the smallest possible length of a broken line AXB (i.e., the union of line segments AX and XB), such that A lies on c_1, B lies on c_2, and X lies on t.*

Solution: Let us consider the circle c_2' with centre O_2' symmetric to c_2 with respect to t. If A, X and B yield the shortest possible broken line as required, the point B' symmetric to B with respect to t must also lie such that the broken line AXB' is the shortest possible with A on c_1, X on t and B' on c_2'. This means that A, X and B' must lie on the line joining O_1 and O_2' as illustrated in the figure below.

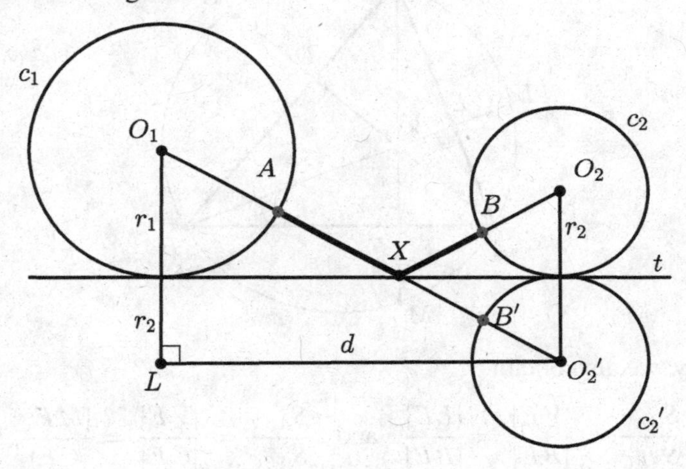

In order to determine its length, we apply the Pythagorean theorem in the right-angled triangle O_1LO_2', with L denoting the point lying both on the line perpendicular to t through O_1 and the line parallel to t through O_2'. Since the lengths of its legs are given by $|O_1L| = r_1 + r_2$ and $|LO_2'| = d$, we see that the smallest length l of the broken line AXB, equal to the smallest length of the broken line AXB', is given by the expression

$$l = |AB'| = \sqrt{(r_1 + r_2)^2 + d^2} - (r_1 + r_2). \qquad \square$$

Problem 143. (B-T-1-10) *Let us consider the triangle ABC with altitudes $h_a = 24$ and $h_b = 32$. Prove that its third altitude h_c fulfils the inequalities*

$$13 < h_c < 96.$$

Solution: For the area S of a triangle ABC, we have

$$S = \tfrac{1}{2} \cdot a \cdot h_a = \tfrac{1}{2} \cdot b \cdot h_b = \tfrac{1}{2} \cdot c \cdot h_c.$$

From this, we obtain

$$a = \frac{2S}{h_a} = \frac{2S}{24}, \quad b = \frac{2S}{h_b} = \frac{2S}{32} \quad \text{and} \quad c = \frac{2S}{h_a},$$

and from the triangle inequalities $|a - b| < c < a + b$, we therefore obtain

$$\frac{2S}{24} - \frac{2S}{32} < \frac{2S}{h_c} < \frac{2S}{24} + \frac{2S}{32}.$$

Dividing by $2S$ therefore yields

$$\frac{1}{24} - \frac{1}{32} < \frac{1}{h_c} < \frac{1}{24} + \frac{1}{32} \iff \frac{1}{96} < \frac{1}{h_c} < \frac{7}{96},$$

and therefore

$$96 > h_c > \frac{96}{7} = 13\frac{5}{7} > 13,$$

which was to be shown. □

Problem 144. (A-T-2-05) *Let* $2s$ *be the length of the perimeter of a triangle* ABC *and let* ρ, r_a, r_b *and* r_c *be the radii of the incircle and three excircles of* ABC, *respectively. Prove that the following inequality holds*:

$$\sqrt{\rho \cdot r_a} + \sqrt{\rho \cdot r_b} + \sqrt{\rho \cdot r_c} \le s.$$

Solution: Let S denote the area of the triangle ABC. By Heron's formula, we have

$$S = \sqrt{s(s-a)(s-b)(s-c)},$$

and the well-known formulas for the inradius and the exradii of a triangle give us

$$\varrho = \frac{S}{s}, \quad r_a = \frac{S}{s-a} \quad r_b = \frac{S}{s-b} \quad r_c = \frac{S}{s-c}.$$

Applying the AM–GM inequality yields

$$\sqrt{\varrho \cdot r_a} = \frac{S}{\sqrt{s(s-a)}} = \sqrt{(s-b)(s-c)} \leq \frac{1}{2}(s-b+s-c),$$
$$\sqrt{\varrho \cdot r_b} = \frac{S}{\sqrt{s(s-b)}} = \sqrt{(s-a)(s-c)} \leq \frac{1}{2}(s-a+s-c),$$
$$\sqrt{\varrho \cdot r_c} = \frac{S}{\sqrt{s(s-c)}} = \sqrt{(s-a)(s-b)} \leq \frac{1}{2}(s-a+s-b),$$

and adding both sides of these three inequalities yields

$$\sqrt{\varrho \cdot r_a} + \sqrt{\varrho \cdot r_b} + \sqrt{\varrho \cdot r_c} \leq \tfrac{1}{2}(6s-4s) = s,$$

as claimed. □

Remark. Equality holds if and only if $s-a = s-b = s-c$, i.e., in an equilateral triangle. □

Problem 145. (B-I-4-03) *We consider a line segment AB of the length c and all right triangles with hypotenuse AB. For all such right triangles, determine the maximum diameter of a circle with the centre on AB which is tangent to the other two sides of the triangle.*

Solution: Let a and b denote the lengths of the legs of a right triangle ABC, and r the radius of a circle as described, whereby M denotes the centre of the circle.

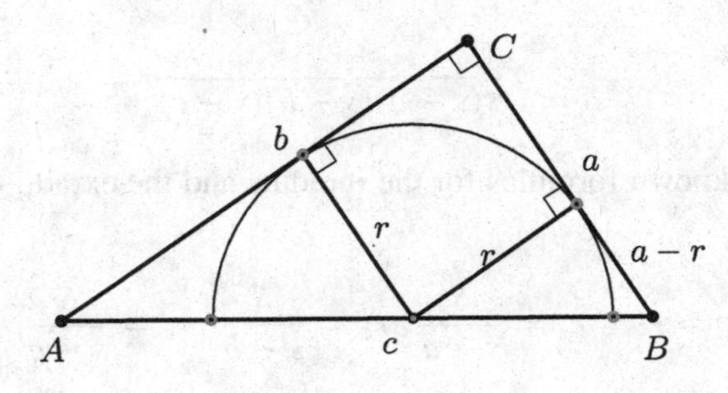

Now let P be the point in which the circle is tangent to BC. Since triangles ABC and MBP are similar, we have

$$\frac{r}{a-r} = \frac{b}{a}, \quad \text{and therefore } r = \frac{ab}{a+b}.$$

We can now apply the HM–QM inequality $H(a, b) \leq Q(a, b)$, where

$$H(a, b) = \frac{2}{\frac{1}{a} + \frac{1}{b}} = \frac{2ab}{a+b} \quad \text{and} \quad Q(a, b) = \sqrt{\frac{a^2 + b^2}{2}} = \sqrt{\frac{c^2}{2}}$$

hold. We see that $d = 2r = H(a, b)$ is bounded from above by the expression $Q(a, b)$, with equality holding only for $a = b$. In this case, the maximal diameter of the circle is given by the expression

$$d = \frac{2a^2}{a+a} = a = \frac{c\sqrt{2}}{2}.$$

□

Another solution: Let C' be the point symmetric to C with respect to AB. We can now consider the quadrilateral $AC'BC$. Let S be the centre of the circle inscribed in the quadrilateral and r its radius.

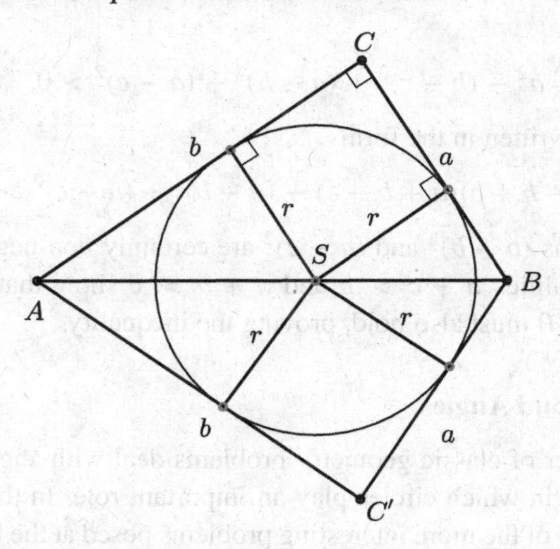

We can calculate the area of the quadrilateral in two different ways.

On the one hand, the area P of $AC'BC$ is the sum of the areas of triangles ABC and $AC'B$. Therefore, we can write

$$P = 2 \cdot \tfrac{1}{2} \cdot ab = ab.$$

On the other hand, the area P of $AC'BC$ is also the sum of the areas of triangles ASC, CSB, BSC' and $C'SA$. Therefore, we can also write

$$P = \tfrac{1}{2}r \cdot b + \tfrac{1}{2}r \cdot a + \tfrac{1}{2}r \cdot a + \tfrac{1}{2}r \cdot b = r \cdot (a + b).$$

Comparing these expressions yields

$$ab = r \cdot (a + b), \quad \text{then } r = \frac{ab}{a + b},$$

and we can complete the proof as in the previous solution. □

Problem 146. (A-I-4-01) *Let a, b, c be the lengths of the sides of a triangle. Prove that the following inequality holds*:

$$3a^2 + 2bc > 2ab + 2ac.$$

Solution: The inequality is equivalent to $3a^2 + 2bc - 2ab - 2ac > 0$, which is in turn equivalent to

$$a^2 - b^2 + 2bc - c^2 + a^2 - 2ab + b^2 + a^2 - 2ac + c^2 > 0$$

or

$$a^2 - (b - c)^2 + (a - b)^2 + (a - c)^2 > 0,$$

which can be written in the form

$$(a - b + c)(a + b - c) + (a - b)^2 + (a - c)^2 > 0.$$

The expressions $(a - b)^2$ and $(a - c)^2$ are certainly non-negative, and the triangle inequalities $a + c > b$ and $a + b > c$ show that $(a - b + c)$ $(a + b - c) > 0$ must also hold, proving the inequality. □

5.2. Circles and Angles

A large number of classic geometry problems deal with angle relations in configurations in which circles play an important role. In this section, we present several of the more interesting problems posed at the Duel from this well-worn path.

PROBLEMS

147. A circle meets each side of a rectangle at two points. Intersection points lying on opposite sides are vertices of two trapezoids. Prove that points of intersection of these two trapezoids lying inside the rectangle are vertices of a cyclic quadrilateral.

148. Let c be a circle with centre O and radius r and l a line containing O. Furthermore, let P and Q be points on c symmetric with respect to l. Also, let X be a point on c such that $OX \perp l$ and A, B be the points of intersection of XP with l and XQ with l, respectively. Prove that $|OA| \cdot |OB| = r^2$ holds.

149. Let O be the circumcentre of an acute-angled triangle ABC. Let D be the foot of the altitude from A to the side BC. Prove that the angle bisector $\angle CAB$ is also the bisector of $\angle DAO$.

150. Two circles $k_1(M_1, r_1)$ and $k_2(M_2, r_2)$ intersect in points S and T. The line M_1M_2 intersects k_1 in points A and B and k_2 in C and D such that B lies in the interior of k_2 and C lies in the interior of k_1. Prove that the lines SC and SB trisect the angle ASD if and only if $|\angle M_1SM_2| = 90°$.

151. We are given a circle c_1 and points X and Y on c_1. Let XY be a diameter of a second circle c_2. We choose a point P on the greater arc XY on c_1 and a point Q on c_2 such that the quadrilateral $PXQY$ is convex and $PX \parallel QY$. Prove that a measure of the angle PYQ is independent of the choice of P (if an appropriate Q exists).

152. Let $ABCD$ be a parallelogram. The circle c with diameter AB passes through the midpoint of the side CD and through the point D. Determine the measure of the angle $\angle ABC$.

153. Let ABC be an isosceles triangle with $|AC| = |BC|$. We are given a circle k with $C \in k$ and AB tangent to k. Furthermore, X is the point of tangency of AB and k, and P and Q are the points of intersection of AC and BC with k, respectively. Prove the following property:

$$|BQ| \cdot |AX|^2 = |AP| \cdot |BX|^2.$$

154. We are given a cyclic quadrilateral $ABCD$ with $\angle BDC = \angle CAD$ and $|AB| = |AD|$. Prove that there exists a circle, which is tangent to all four sides of the quadrilateral $ABCD$.

SOLUTIONS

Problem 147. (B-I-3-16) *A circle meets each side of a rectangle at two points. Intersection points lying on opposite sides are vertices of two trapezoids. Prove that points of intersection of these two trapezoids lying inside the rectangle are vertices of a cyclic quadrilateral.*

Solution: First of all, it is clear that both trapezoids must be equilateral. Let α and β denote the angles of one trapezoid with ($\alpha + \beta = 180°$) and γ and δ the angles of the other (with $\gamma + \delta = 180°$), as shown in the picture below.

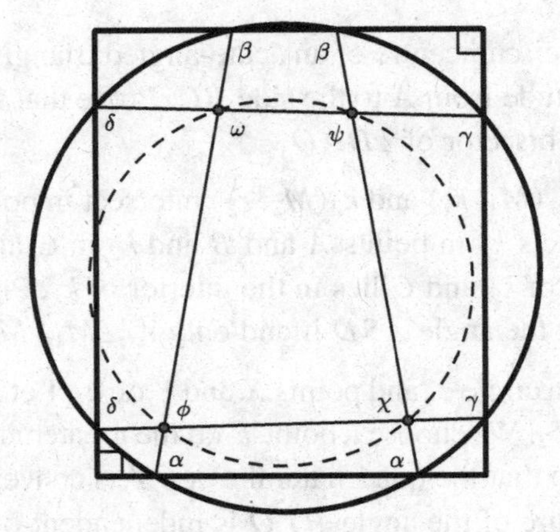

One can now easily compute all the angles of the quadrilateral in question:

$$\phi = 360° - 90° - (180° - \alpha) - (180° - \delta) = \alpha + \delta - 90°,$$
$$\chi = \alpha + \gamma - 90°,$$
$$\psi = \beta + \gamma - 90°,$$
$$\omega = \beta + \delta - 90°.$$

We therefore have

$$\phi + \psi = (\alpha + \delta - 90°) + (\beta + \gamma - 90°) = (\alpha + \beta) + (\gamma + \delta) - 180° = 180°,$$

which proves our claim. □

Problem 148. (A-T-2-11) *Let c be a circle with centre O and radius r and l a line containing O. Furthermore, let P and Q be points on c symmetric* *with respect to l. Also, let X be a point on c such that $OX \perp l$ and A, B be the points of intersection of XP with l and XQ with l, respectively. Prove that $|OA| \cdot |OB| = r^2$ holds.*

Solution: We name $Y = PQ \cap \ell$ and $|PY| = |QY| = x$. Since $PQ \perp \ell$, the triangles AXO and APY are similar and we have

$$\frac{|OA|}{|OA| - |OY|} = \frac{r}{x} \Rightarrow x \cdot |OA| = r \cdot |OA| - r \cdot |OY|$$

and thus

$$|OA| = \frac{r \cdot |OY|}{r - x}.$$

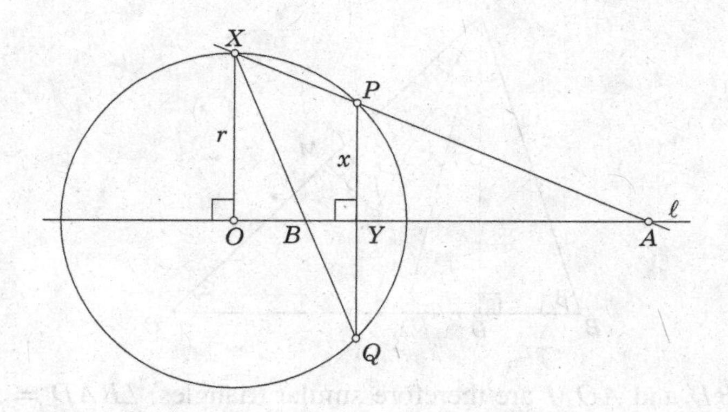

Analogously, since the triangles BXO and BQY are similar, we have

$$\frac{|OB|}{|OY| - |OB|} = \frac{r}{x} \Rightarrow x \cdot |OB| = r \cdot |OY| - r \cdot |OB|$$

and

$$|OB| = \frac{r \cdot |OY|}{r + x}.$$

It therefore follows that

$$|OA| \cdot |OB| = \frac{r^2 \cdot |OY|^2}{(r - x)(r + x)} = \frac{r^2(r^2 - x^2)}{r^2 - x^2} = r^2$$

holds, as claimed. □

Problem 149. (A-I-3-13) *Let O be the circumcentre of an acute-angled triangle ABC. Let D be the foot of the altitude from A to the side BC. Prove that the angle bisector $\angle CAB$ is also the bisector of $\angle DAO$.*

Solution: Without loss of generality, we can assume that $\beta \geq \gamma$ holds, as illustrated in the picture below. Let M be the midpoint of the side AC of the triangle ABC and U the point of intersection of the angle bisector at A with the side BC. We see that the angle $\angle ABC$ is equal to half of the angle $\angle AOC$ in the circumcircle of ABC, and therefore

$$|\angle AOM| = |\angle ABD| = |\angle ABC| = \beta$$

holds.

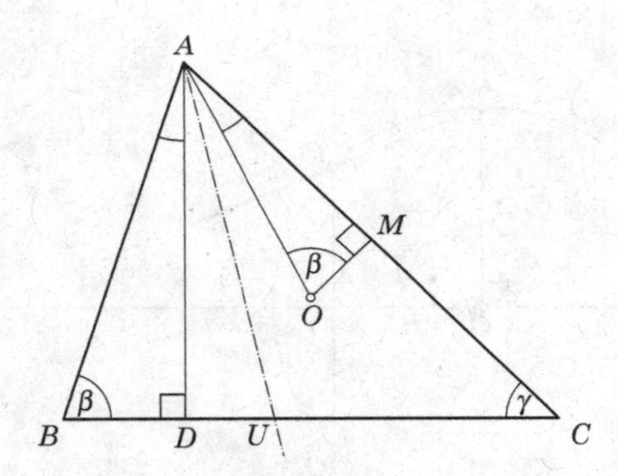

Since ABD and AOM are therefore similar triangles, $\angle BAD = \angle OAM$ follows, and we therefore have $\angle DAU = \angle OAU$, as claimed. □

Problem 150. (C-T-2-14) *Two circles $k_1(M_1, r_1)$ and $k_2(M_2, r_2)$ intersect in points S and T. The line M_1M_2 intersects k_1 in points A and B and k_2 in C and D such that B lies in the interior of k_2 and C lies in the interior of k_1. Prove that the lines SC and SB trisect the angle ASD if and only if $|\angle M_1SM_2| = 90°$.*

Solution: Naming $\angle SAB = \alpha$ and $\angle SDC = \beta$, we have $\angle SBA = 90° - \alpha$, and therefore $\angle BSD = 90° - \alpha - \beta$. On the other hand, we also have $\angle SCD = 90° - \beta$, and therefore $\angle CSA = 90° - \alpha - \beta$, and these two angles are therefore certainly equal. Since $\angle ASB = 180° - \alpha - \beta$, we have $\angle CSB = \alpha + \beta$.

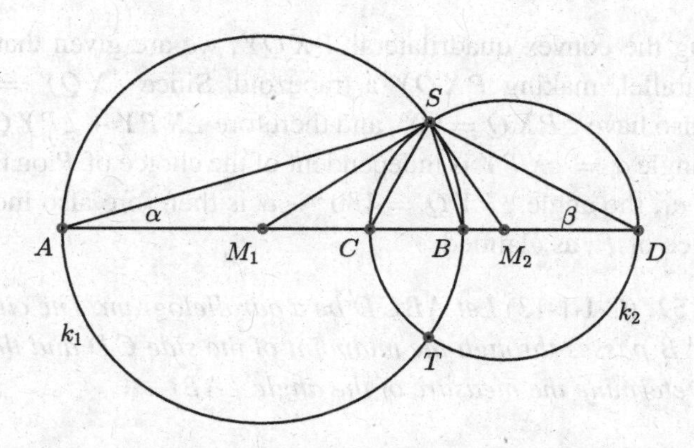

We are now ready to consider the circumstances under which SC and SB trisect $\angle ASD$. This is equivalent to $90° - \alpha - \beta = \alpha + \beta$, which is itself equivalent to $\alpha + \beta = 45°$. Since $\angle SM_1B = 2\alpha$ and $\angle SM_2C = 2\beta$, this is equivalent to

$$\angle SM_1B + \angle SM_2C = \angle SM_1M_2 + \angle SM_2M_1 = 2(\alpha + \beta) = 90°,$$

and since $\angle M_1SM_2 = 180° - (\angle SM_1M_2 + \angle SM_2M_1)$, this is equivalent to $\angle M_1SM_2 = 90°$, as claimed. □

Problem 151. (A-I-3-10) *We are given a circle c_1 and points X and Y on c_1. Let XY be a diameter of a second circle c_2. We choose a point P on the greater arc XY on c_1 and a point Q on c_2 such that the quadrilateral $PXQY$ is convex and $PX \parallel QY$. Prove that the size of the angle PYQ is independent of the choice of P (if an appropriate Q exists).*

Solution: As we can see in the figure, we certainly have $\angle XQY = 90°$, since XY is a diameter of c_2 and Q lies on c_2.

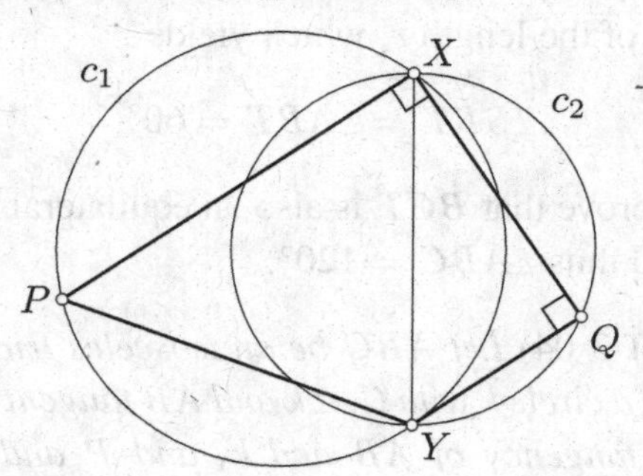

Considering the convex quadrilateral $PXQY$, we are given that PX and QY are parallel, making $PXQY$ a trapezoid. Since $\angle XQY = 90°$, we therefore also have $\angle PXQ = 90°$, and therefore $\angle XPY + \angle PYQ = 180°$. Since the angle $\alpha = \angle XPY$ is independent of the choice of P on the greater arc XY of c_1, the angle $\angle PYQ = 180° - \alpha$ is therefore also independent of the choice of P, as claimed. □

Problem 152. (C-I-1-13) *Let $ABCD$ be a parallelogram. The circle c with diameter AB passes through the midpoint of the side CD and through the point D. Determine the measure of the angle $\angle ABC$.*

Solution: Let S and T denote the midpoints of the sides AB and CD of the given parallelogram respectively, and $2r$ their lengths, as illustrated in the picture below. Since the points D and T lie on the circle with diameter AB, we have

$$|SA| = |SB| = |ST| = |AD| = |BC| = r.$$

Furthermore, since DT lies on the side CD of the parallelogram $ABCD$ opposite AB, the chords AB and DT of the circle c are parallel.

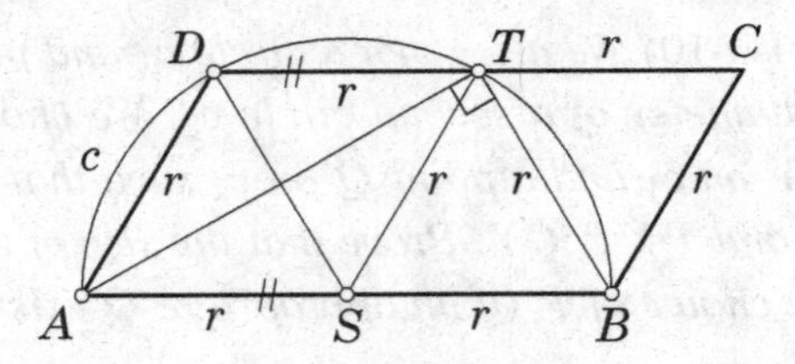

The quadrilateral $ABTD$ is therefore an isosceles trapezoid with bases AB and DT and with $|AD| = |BT| = r$. SBT is therefore an equilateral triangle with sides of the length r, which yields

$$\angle SBT = \angle ABT = 60°.$$

Similarly, we can prove that BCT is also an equilateral triangle with sides of the length r, and thus $\angle ABC = 120°$. □

Problem 153. (B-T-3-04) *Let ABC be an isosceles triangle with $|AC| = |BC|$. We are given a circle k with $C \in k$ and AB tangent to k. Furthermore, X is the point of tangency of AB and k, and P and Q are the points* *of intersection of AC and BC with k, respectively. Prove the following property*:

$$|BQ| \cdot |AX|^2 = |AP| \cdot |BX|^2.$$

Solution: The power of the point A with respect to the circle k yields

$$|AX|^2 = |AP| \cdot |AC| \tag{1}$$

and similarly, from the power of the point B with respect to the same circle k we get

$$|BX|^2 = |BQ| \cdot |BC| = |BQ| \cdot |AC|. \tag{2}$$

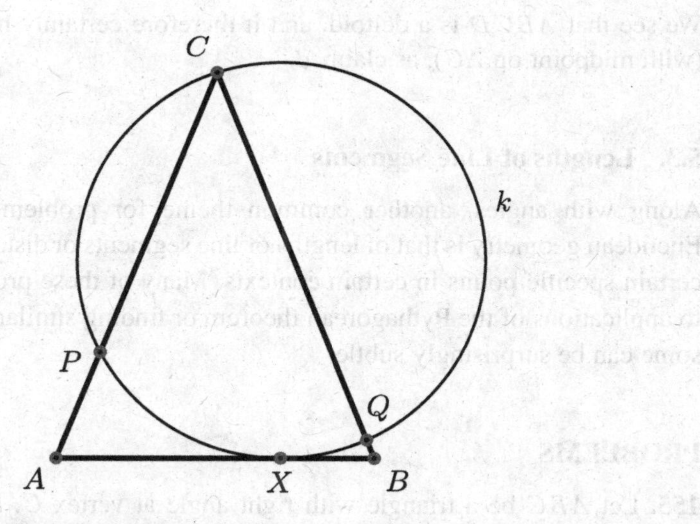

Eliminating $|AC|$ from (1) and (2), we therefore obtain

$$\frac{|AX|^2}{|AP|} = \frac{|BX|^2}{|BQ|}.$$

which is equivalent to the claimed property, concluding the proof. □

Problem 154. (A-I-2-12) *We are given a cyclic quadrilateral $ABCD$ with $\angle BDC = \angle CAD$ and $|AB| = |AD|$. Prove that there exists a circle, which is tangent to all four sides of the quadrilateral $ABCD$.*

Solution: Since $\angle CAD = \angle BDC = \angle BAC$, we have $|BC| = |CD|$. It therefore follows that triangles ABC and ADC are congruent, since we have $|AB| = |AD|$, $|BC| = |BD|$ and AC is a common side.

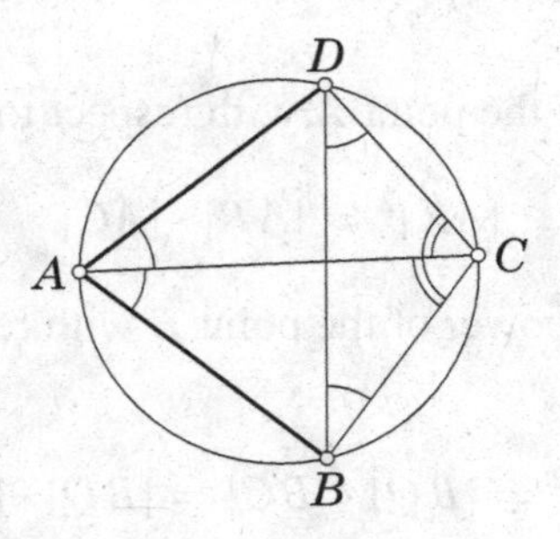

We see that $ABCD$ is a deltoid, and it therefore certainly has an incircle (with midpoint on AC), as claimed. □

5.3. Lengths of Line Segments

Along with angles, another common theme for problems in classical Euclidean geometry is that of lengths of line segments or distances between certain specific points in certain contexts. Many of these problems reduce to applications of the Pythagorean theorem or finding similar triangles, but some can be surprisingly subtle.

PROBLEMS

155. Let ABC be a triangle with right angle at vertex C. Let ACP and BCQ be right isosceles triangles external to ABC with right angles at P and Q, respectively. Furthermore, let F be the perpendicular foot of C on AB, and D and E be the points of intersection of the line AC with PF and the line BC with QF, respectively. Prove that $|DC| = |EC|$.

156. We are given a trapezoid $ABCD$ with $AB \parallel CD$ and $|AB| = 2|CD|$. Let M be the common point of the diagonals AC and BD and E the midpoint of AD. Lines EM and CD intersect in P. Prove that $|CP| = |CD|$ holds.

157. We are given bases $|AB| = 23$ and $|CD| = 5$ of a trapezoid $ABCD$ with diagonals $|AC| = 25$ and $|BD| = 17$. Determine the lengths of its sides BC and AD.

SOLUTIONS

Problem 155. (B-I-2-15) *Let ABC be a triangle with right angle at vertex C. Let ACP and BCQ be right isosceles triangles external to ABC with right angles at P and Q, respectively. Furthermore, let F be the perpendicular foot of C on AB, and D and E be the points of intersection of the line AC with PF and the line BC with QF, respectively. Prove that $|DC| = |EC|$.*

Solution: Since AP and AF are perpendicular to CP and CF, respectively, P and F are points of the circle k_1 with diameter AC. The triangle ACP is external to ABC, and $AFCP$ is therefore a convex cyclic quadrilateral with circumcircle k_1. Consequently, since ACP is an isosceles right triangle, we have $\angle CFP = \angle CAP = 45°$. Analogously, $BQCF$ is a cyclic quadrilateral with circumcircle k_2, and $\angle QFC = \angle QBC = 45°$.

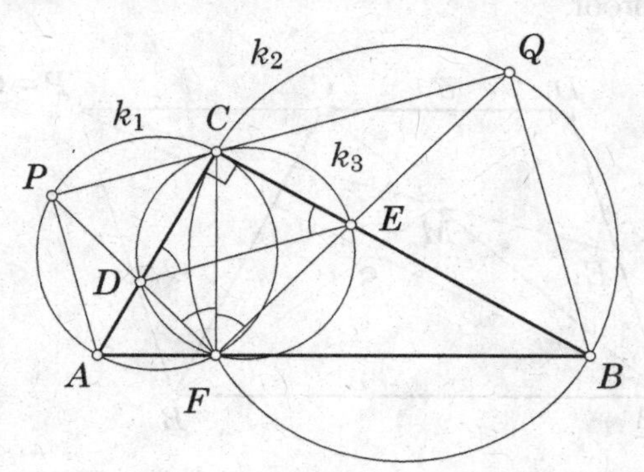

This means that $\angle QFP = \angle EFD = 90° = \angle DCE$, and $CDEF$ is therefore also a cyclic quadrilateral with DE as a diagonal. From this, we see that $\angle EDC = \angle EFC = 45° = \angle CFD = \angle CED$ holds, and CDE is therefore an isosceles right triangle with $|DC| = |EC|$, as claimed. □

Problem 156. (A-I-2-11) *We are given a trapezoid $ABCD$ with $AB \parallel CD$ and $|AB| = 2|CD|$. Let M be the common point of the diagonals AC and*

BD and E the midpoint of AD. Lines EM and CD intersect in P. Prove that $|CP| = |CD|$ holds.

Solution: Let S be the midpoint of the diagonal BD. Furthermore, let Q be the point of intersection of the line AS with CD. Q is then a vertex of a parallelogram, namely $ABQD$, as illustrated in the figure below. Since $|AB| = 2\,|CD|$ is given, C is the midpoint of the side DQ of this parallelogram.

From the similarity of the triangles ABM and CDM, we obtain

$$|BM|{:}|DM| = |AB|{:}|CD| = 2{:}1,$$

and since S is the midpoint of BD, we also have

$$|BM|{:}|DM| = |DM|{:}|MS| = 2{:}1.$$

The common point M of the diagonals AC and BD of the given trapezoid $ABCD$ must therefore simultaneously be the centroid of the triangle AQD. Its median EQ is therefore identical with EM, and this implies $Q = P$, completing the proof. □

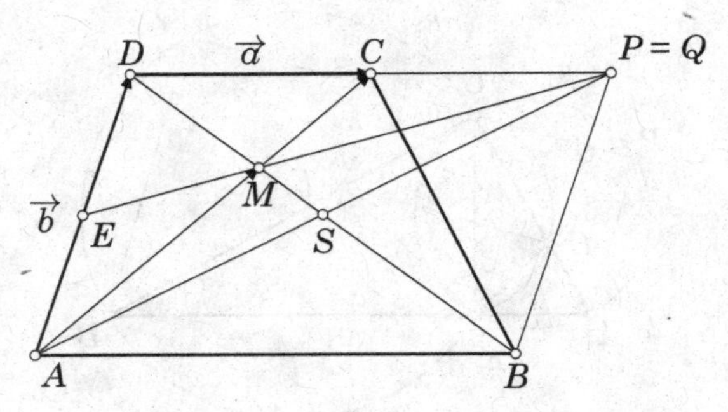

Another solution: Let $\overrightarrow{DC} = \vec{a}$ and $\overrightarrow{AD} = \vec{b}$. We then have $\overrightarrow{AE} = \frac{1}{2}\vec{b}$ and $\overrightarrow{AM} = \frac{2}{3}(\vec{a} + \vec{b})$, since triangles MAB and MCD are similar with ratio 2:1. We therefore have $\overrightarrow{EM} = \frac{2}{3}\vec{a} + \frac{1}{6}\vec{b}$.

The vector $\overrightarrow{DP}$ can now be written in two ways, and we have

$$-\tfrac{1}{2}\vec{b} + \lambda\left(\tfrac{2}{3}\vec{a} + \tfrac{1}{6}\vec{b}\right) = \mu\vec{a}.$$

Comparing coefficients yields $\lambda = 3$, and thus $\mu = 2$, and we see that DP is twice as long as DC, as claimed. □

Problem 157. (A-I-2-10) *We are given bases* $|AB| = 23$ *and* $|CD| = 5$ *of a trapezoid* $ABCD$ *with diagonals* $|AC| = 25$ *and* $|BD| = 17$. *Determine the lengths of its sides* BC *and* AD.

Solution: Let E be the point on the line AB such that $DB \parallel CE$. We can calculate the area P of the triangle AEC from Heron's formula

$$P = \sqrt{s(s-a)(s-b)(s-c)},$$

where $a = |AE| = 23 + 5 = 28$, $b = |AC| = 25$, and $c = |CE| = |DB| = 17$ are the lengths of the sides of the triangle and

$$s = \frac{a+b+c}{2}.$$

Since $s = \frac{1}{2}(28 + 25 + 17) = 35$, $s - a = 7$, $s - b = 10$ and $s - c = 18$ hold, we obtain

$$P = \sqrt{35 \cdot 7 \cdot 10 \cdot 18} = 7 \cdot 5 \cdot 6 = 210.$$

On the other hand, we can also write

$$P = \tfrac{1}{2} \cdot |AE| \cdot h = \tfrac{1}{2} \cdot 28 \cdot h = 14h,$$

where h is the altitude of the triangle AEC (and therefore also of the trapezoid $ABCD$). This gives us $14h = 210$, and therefore $h = 15$.

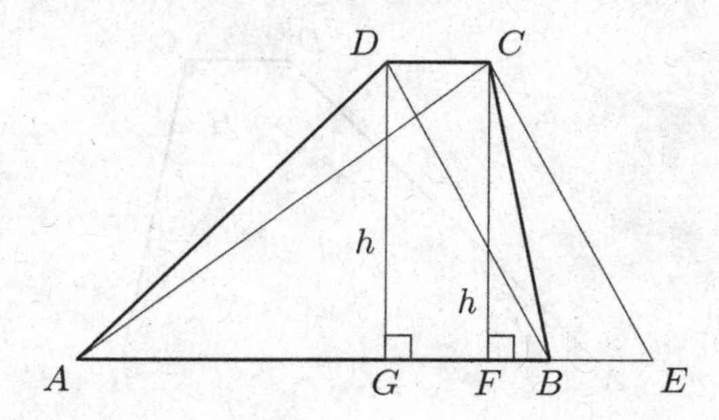

Let us now consider the triangle ABC with the altitude $|CF| = 15$. From the Pythagorean theorem in the triangle AFC, we obtain

$$|AF| = \sqrt{|AC|^2 - |CF|^2} = \sqrt{625 - 225} = \sqrt{400} = 20.$$

We therefore have $|FB| = |AB| - |AF| = 23 - 20 = 3$, and the Pythagorean theorem in triangle FBC then yields

$$|CB| = \sqrt{|CF|^2 + |FB|^2} = \sqrt{15^2 + 3^2} = \sqrt{225 + 9} = \sqrt{234} = 3\sqrt{26}.$$

Now let us consider the triangle ABD with the altitude $|DG| = 15$. From the Pythagorean theorem in the triangle GBD, we obtain

$$|BG| = \sqrt{|BD|^2 - |DG|^2} = \sqrt{289 - 225} = \sqrt{64} = 8.$$

We therefore have $|AG| = |AB| - |GB| = 23 - 8 = 15$, and the Pythagorean theorem in the triangle AGD then yields

$$|AD| = \sqrt{|DG|^2 + |AG|^2} = \sqrt{15^2 + 15^2} = \sqrt{2 \cdot 15^2} = 15\sqrt{2}.$$

Another solution: Let H denote the common point of the segments AC and BD and ω the angle $\angle AHB$, as illustrated in the figure below. The lines AB and CD are parallel, and it follows that the triangles AHB and CHD are similar with coefficient $|AB|:|CD| = 23:5$. This yields

$$|AH| = 25 \cdot \tfrac{23}{28}, \quad |BH| = 17 \cdot \tfrac{23}{28}, \quad |CH| = 25 \cdot \tfrac{5}{28}, \quad \text{and } |DH| = 17 \cdot \tfrac{5}{28}.$$

Applying the cosine rule to triangle ABH, we obtain

$$\cos \omega = \frac{|AH|^2 + |BH|^2 - |AB|^2}{2 \cdot |AH| \cdot |BH|} = \frac{13}{85}.$$

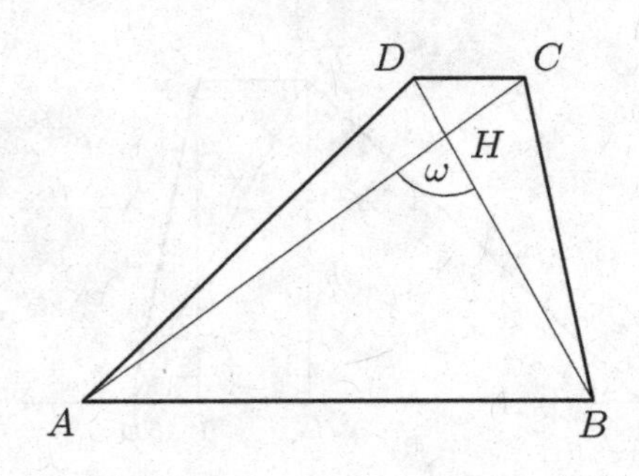

For the supplementary angles to ω, we therefore have

$$\cos \angle AHD = \cos \angle BHC = \cos(180^\circ - \omega) = -\cos\omega = -\tfrac{13}{85}.$$

Now, applying the cosine rule in triangle BHC yields

$$|BC|^2 = |BH|^2 + |CH|^2 - 2\,|BH| \cdot |CH| \cos \angle BHC = 234 = 3^2 \cdot 26$$

and in triangle AHD,

$$|AD|^2 = |AH|^2 + |DH|^2 - 2\,|AH| \cdot |DH| \cos \angle AHD = 450 = 15^2 \cdot 2.$$

We therefore have $|BC| = 3\sqrt{26}$ and $|AD| = 15\sqrt{2}$. □

5.4. Triangles

Close relatives of line segment problems are those relating to triangles. These often ask questions concerning the various triangle centres, such as the orthocentre or the circumcentre, to name just two.

PROBLEMS

158. The orthocentre H of an acute-angled triangle ABC is reflected on the sides a, b and c yielding points A_1, B_1 and C_1, respectively. We are given $\angle C_1AB_1 = \angle CA_1B$, $\angle A_1BC_1 = \angle AB_1C$ and $\angle B_1CA_1 = \angle BC_1A$. Prove that ABC must be an equilateral triangle.

159. Two lines p and q intersect in a point V. The line p is tangent to a circle k in the point A. The line q intersects k in the points B and C. The angle bisector of $\angle AVB$ intersects the segments AB and AC in the points K and L, respectively. Prove that the triangle KLA is isosceles.

160. Let D be a point on the side BC of a given triangle ABC such that

$$|AB| + |BD| = |AC| + |CD|$$

holds. The line segment AD intersects the incircle of ABC at X and Y with X closer to A. Let E be the point of tangency of the incircle of ABC with BC. Let I denote the incentre of ABC and M the midpoint of the line segment BC. Show that

(a) the line EY is perpendicular to AD and
(b) $|XD| = 2|IM|$ holds.

SOLUTIONS

Problem 158. (A-I-2-08) *The orthocentre H of an acute-angled triangle ABC is reflected on the sides a, b and c yielding points A_1, B_1 and C_1, respectively. We are given $\angle C_1AB_1 = \angle CA_1B$, $\angle A_1BC_1 = \angle AB_1C$ and $\angle B_1CA_1 = \angle BC_1A$. Prove that ABC must be an equilateral triangle.*

Solution: Naming $\angle CAB = \alpha$, $\angle ABC = \beta$ and $\angle BCA = \gamma$, we first note that

$$\begin{aligned}\angle BA_1C &= \angle BHC \\ &= 180^\circ - \angle HBC - \angle HCB \\ &= 180^\circ - (90^\circ - \gamma) - (90^\circ - \beta) \\ &= \beta + \gamma\end{aligned}$$

holds. Since $\alpha = 180^\circ - (\beta + \gamma)$, A_1 must lie on the circumcircle of ABC. Similarly, B_1 and C_1 must also lie on the circumcircle of ABC, and we therefore have $\angle CAB_1 = \angle CBB_1 = 90^\circ - \gamma$ and $\angle BAC_1 = \angle BCC_1 = 90^\circ - \beta$. It therefore follows that $\angle C_1AB_1 = \alpha + (90^\circ - \gamma) + (90^\circ - \beta) = 2\alpha$ holds, and since $\angle C_1AB_1 = \angle CA_1B$, we therefore have $2\alpha = \beta + \gamma = 180^\circ - \alpha$, and therefore $\alpha = 60^\circ$.

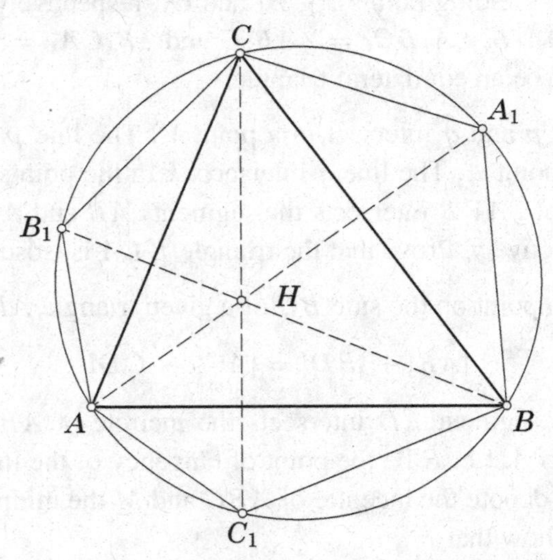

Since the same holds for β and γ, the triangle ABC must be equilateral. □

Problem 159. (B-T-1-13) *Two lines p and q intersect in a point V. The line p is tangent to a circle k in the point A. The line q intersects k in the points B and C. The angle bisector of $\angle AVB$ intersects the segments AB and AC in the points K and L, respectively. Prove that the triangle KLA is isosceles.*

Solution: Since p is tangent to the circle k, $\angle VAB = \angle VCA$ must hold.

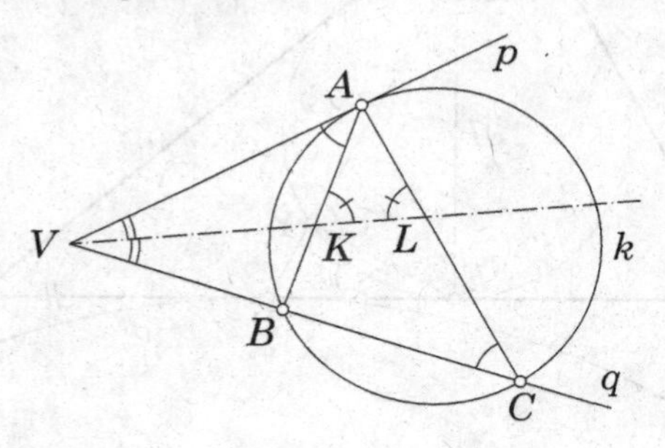

Since KL is the angle bisector of $\angle AVC$, we have

$$\angle AKL = \angle VAK + \angle AVK = \angle VAB + \angle AVK$$
$$= \angle VCA + \angle CVL = \angle VCL + \angle CVL = \angle ALV = \angle ALK.$$

This means that AKL is an isosceles triangle with the base KL, which completes the proof. □

Problem 160. (A-T-2-08) *Let D be a point on the side BC of a given triangle ABC such that*

$$|AB| + |BD| = |AC| + |CD|$$

holds. The line segment AD intersects the incircle of ABC at X and Y with X closer to A. Let E be the point of tangency of the incircle of ABC with BC. Let I denote the incentre of ABC and M the midpoint of the line segment BC. Show that

(a) *the line EY is perpendicular to AD and*
(b) *$|XD| = 2|IM|$ holds.*

Solution: (a) The homothety mapping the excircle at the side BC of triangle ABC (tangent to BC at the point D) to the incircle of the same triangle maps the point D to X. The tangent of the incircle in X is therefore parallel to BC. It follows that XE is a diameter of the incircle and thus XYE is a right angle, as claimed.

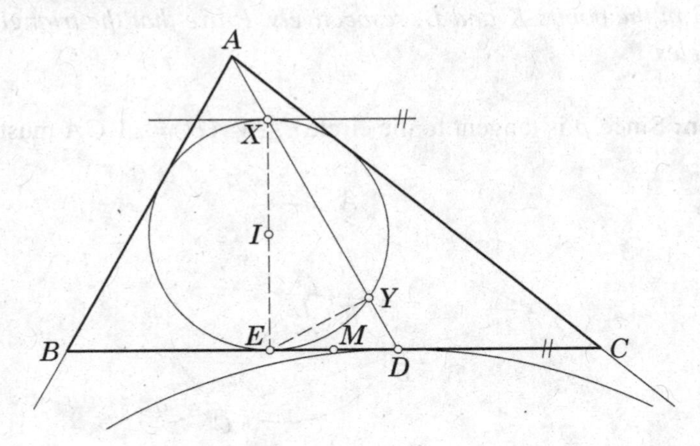

(b) It is well known (and verified by some easy computation) that $|BE| = |DC| = s - b$ holds, with $2s = a + b + c$. M is therefore the midpoint of the segment ED. Since I is also the midpoint of XE, triangles EMI and EDX are similar, and we therefore also have $|XD| = 2\,|IM|$, as claimed. □

5.5. Area and Perimeter

In this section, we focus our attention on problems pertaining to the areas or perimeters of certain plane figures.

PROBLEMS

161. We are given an isosceles right triangle EBC with the right angle at C and $|BC| = 2$. Determine all possible values of the area of a trapezoid $ABCD$ ($AB \parallel CD$) in which E is the midpoint of AD.

162. We are given a square with the area P and circumcircle k. Let us consider two lines parallel to the sides of the square and passing through a common interior point of the square. These parallels intersect k in the vertices of a quadrilateral with the area Q. Prove that $P \geq Q$ must hold.

163. Let us consider a trapezoid with sides of lengths 3, 3, 3, k, with positive integer k. Determine the maximum area of such a trapezoid.

164. Let us consider a unit square $ABCD$. On its sides BC and CD, determine points E and F respectively, such that $|BE| = |DF|$ holds, and the triangles ABE and AEF have the same perimeter.

165. We are given a triangle ABC. Prove that for any triple u, v, w of positive real numbers, there exists a point P inside the triangle ABC such that

$$S_{ABP}:S_{BCP}:S_{CAP} = u:v:w.$$

(Note that S_{XYZ} denotes the area of a triangle XYZ.)

166. We are given a regular hexagon $ABCDEF$ with the area P. Lines CD and EF intersect at the point G. Determine the areas of the triangles ABG and BCG in terms of P.

SOLUTIONS

Problem 161. (B-I-2-11) *We are given an isosceles right triangle EBC with the right angle at C and $|BC| = 2$. Determine all possible values of the area of a trapezoid $ABCD$ ($AB \parallel CD$) in which E is the midpoint of AD.*

Solution: Let F be the midpoint of the side BC in such a trapezoid $ABCD$. Let us consider the right-angled triangle EFC.

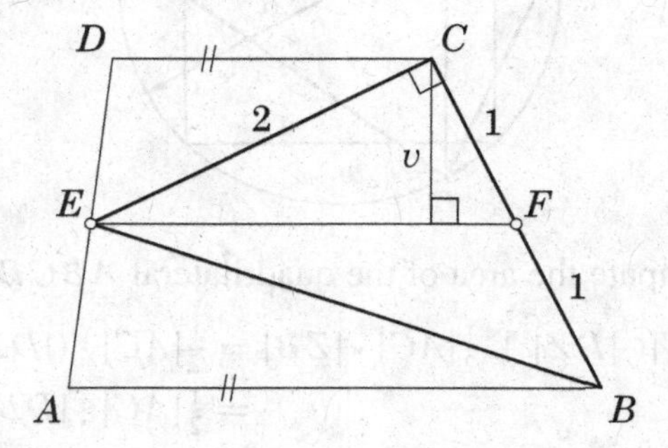

Applying the Pythagorean theorem gives us $|EF| = \sqrt{5}$. Letting v denote the altitude of the triangle (as shown in the figure), we can calculate its (double) area in two ways, which yields

$$\sqrt{5} \cdot v = |EF| \cdot v = |EC| \cdot |FC| = 2 \cdot 1 = 2.$$

From this equation, we obtain

$$v = \frac{2\sqrt{5}}{5},$$

and the area P of the trapezoid $ABCD$ is therefore equal to

$$P = |EF| \cdot 2v = \sqrt{5} \cdot 2 \cdot \tfrac{2\sqrt{5}}{5} = 4. \qquad \square$$

Problem 162. (B-I-2-07) *We are given a square with the area P and circumcircle k. Let us consider two lines parallel to the sides of the square and passing through a common interior point of the square. These parallels intersect k in the vertices of a quadrilateral with the area Q. Prove that $P \geq Q$ must hold.*

Solution: Let R be the radius of the circle k and a be the length of a side of the square inscribed in k. We then have $a = R\sqrt{2}$ and $P = 2R^2$.

Let Z be an interior point of the square. Let the two lines parallel to the sides of the square and passing through Z intersect k in points A, C and B, D, respectively.

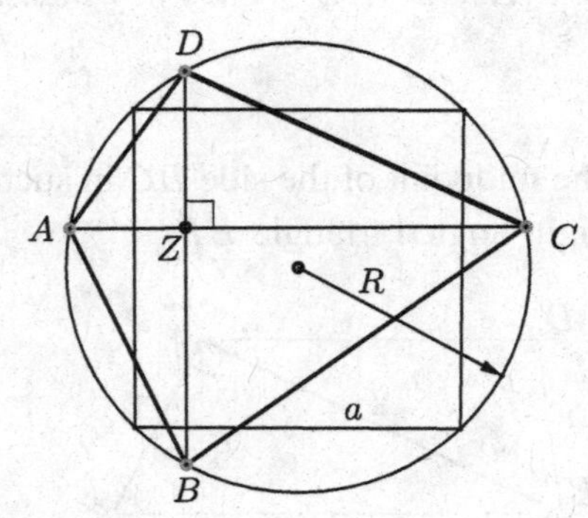

We can now compute the area of the quadrilateral $ABCD$:

$$Q = \tfrac{1}{2}|AC| \cdot |DZ| + \tfrac{1}{2}|AC| \cdot |ZB| = \tfrac{1}{2}|AC| \cdot (|DZ| + |ZB|)$$
$$= \tfrac{1}{2}|AC| \cdot |DB|.$$

Because $|AC| \leq 2R$ and $|BD| \leq 2R$, we have

$$Q = \tfrac{1}{2}|AC| \cdot |BD| \leq \tfrac{1}{2} \cdot 2R \cdot 2R = 2R^2 = P$$

and the inequality $Q \leq P$ holds, as claimed.

Remark 1. If the point Z is any interior point of the circle, the inequality $P \geq Q$ is also true.

Remark 2. In the inequality $P \geq Q$, equality holds if and only if Z is the centre of the circle k. □

Problem 163. (C-I-3-10) *Let us consider a trapezoid with sides of lengths* $3, 3, 3, k$, *with positive integer* k. *Determine the maximum area of such a trapezoid.*

Solution: For such a trapezoid to exist, the inequalities $1 \leq k \leq 8$ must be fulfilled. Let P_k denote the area of the trapezoid with one of side length k.

We note that the altitudes of the trapezoids for $k = 2$ and $k = 4$ are the same, and therefore $P_2 < P_4$ must hold. Similarly, the altitudes of the trapezoids for $k = 1$ and $k = 5$ are the same, and $P_1 < P_5$ must hold.

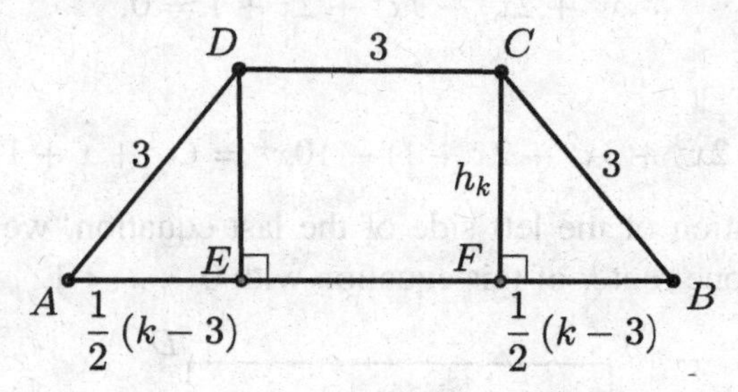

For $3 \leq k \leq 8$, we have $|AE| = |FB| = \frac{1}{2}(k-3)$. For the lengths of their altitudes h_k we then have

$$h_k = \sqrt{BC^2 - FB^2} = \sqrt{9 - \frac{(k-3)^2}{4}},$$

and for the corresponding areas we have

$$P_k = \frac{k+3}{2} \cdot \sqrt{9 - \frac{(k-3)^2}{4}}.$$

Computing the areas P_k of the trapezoids for $k = 3, 4, 5, 6, 7, 8$, we therefore obtain

$$P_3 = \sqrt{81} = \frac{\sqrt{1296}}{4}, \quad P_4 = \frac{\sqrt{1715}}{4}, \quad P_5 = \sqrt{128} = \frac{\sqrt{2048}}{4},$$
$$P_6 = \frac{\sqrt{2187}}{4}, \quad P_7 = \sqrt{125} = \frac{\sqrt{2000}}{4}, \quad P_8 = \frac{\sqrt{1331}}{4}.$$

For $k = 6$ the area of the trapezoid is therefore maximal, with $P_6 = \frac{\sqrt{2187}}{4}$. □

Problem 164. (A-T-2-10) *Let us consider a unit square $ABCD$. On its sides BC and CD, determine points E and F respectively, such that $|BE| = |DF|$ holds, and the triangles ABE and AEF have the same perimeter.*

Solution: Let $|BE| = |DF| = x$, and thus $|CE| = |CF| = 1 - x$. Using conditions from the given problem we see that the equation

$$1 + x + \sqrt{1 + x^2} = 2\sqrt{1 + x^2} + \sqrt{2\,(1 - x)^2}$$

holds. This is equivalent to

$$x^4 + 2x^3 - 7x^2 + 2x + 1 = 0,$$

or

$$0 = (x^4 + 2x^3 + 3x^2 + 2x + 1) - 10x^2 = (x^2 + x + 1)^2 - 10x^2.$$

After factorization of the left side of the last equation, we see that there exists exactly one root x of this equation with $0 < x < 1$.

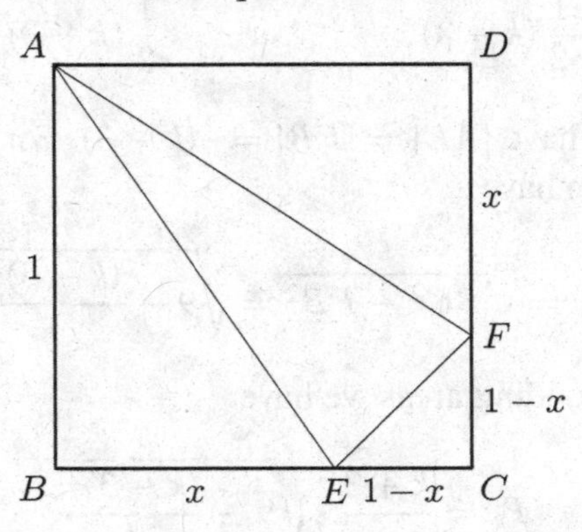

The solution of the given problem is therefore

$$x = \tfrac{1}{2}(\sqrt{10} - 1 - \sqrt{7 - 2\sqrt{10}}).\qquad \square$$

Problem 165. (A-T-2-15) *We are given a triangle ABC. Prove that for any triple u, v, w of positive real numbers, there exists a point P inside the triangle ABC such that*

$$S_{ABP}:S_{BCP}:S_{CAP} = u:v:w.$$

(*Note that S_{XYZ} denotes the area of a triangle XYZ.*)

Solution: First of all, we note that if the point D divides AC in the ratio $u:v$,

$$\frac{S_{ABP}}{S_{BCP}} = u:v$$

holds for every point P on BD.

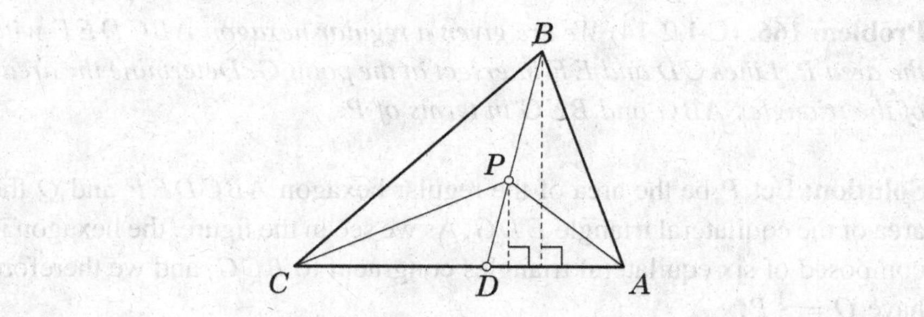

This can be seen in the following way. The triangles ABD and BCD have a common altitude, as do the triangles ADP and CDP. Thus, the ratio of their areas is equal to

$$\frac{S_{ABD}}{S_{BCD}} = \frac{S_{ADP}}{S_{CDP}} = |AD|:|DC| = \frac{u}{v}.$$

It is well known (and can easily be verified by simply multiplying) that $a:b = c:d$ implies

$$a:b = (a+c):(b+d) = (a-c):(b-d).$$

It therefore follows that

$$\frac{S_{ABP}}{S_{BCP}} = \frac{S_{ABD} - S_{ADP}}{S_{BCD} - S_{CDP}} = \frac{u}{v}$$

indeed holds. Now, let $|BP|:|PD| = x:y$. We then have

$$\frac{S_{ADP}}{S_{ABP}} = \frac{y}{x} \quad \text{and} \quad \frac{S_{CDP}}{S_{BCP}} = \frac{y}{x},$$

and can therefore compute the ratio

$$\frac{S_{ACP}}{S_{BCP}} = \frac{S_{ADP} + S_{CDP}}{S_{BCP}} = \frac{S_{ADP}}{S_{BCP}} + \frac{S_{CDP}}{S_{BCP}}$$

$$= \frac{S_{ADP}}{S_{ABP}} \cdot \frac{S_{ABP}}{S_{BCP}} + \frac{y}{x} = \frac{y}{x} \cdot \frac{u}{v} + \frac{y}{x} = \frac{y}{x} \cdot \left(\frac{u+v}{v}\right).$$

We wish to show that P can be chosen in such a way that this ratio is equal to $w:v$. This can be achieved by selecting $y = \frac{w}{v}$, $x = \frac{v}{u+v}$. We have therefore shown that a selection of P fulfilling the required condition is possible for all positive values of u, v and w. □

Problem 166. (C-I-2-14) *We are given a regular hexagon $ABCDEF$ with the area P. Lines CD and EF intersect at the point G. Determine the areas of the triangles ABG and BCG in terms of P.*

Solution: Let P be the area of the regular hexagon $ABCDEF$ and Q the area of the equilateral triangle EDG. As we see in the figure, the hexagon is composed of six equilateral triangles congruent to BCG, and we therefore have $Q = \frac{1}{6}P$.

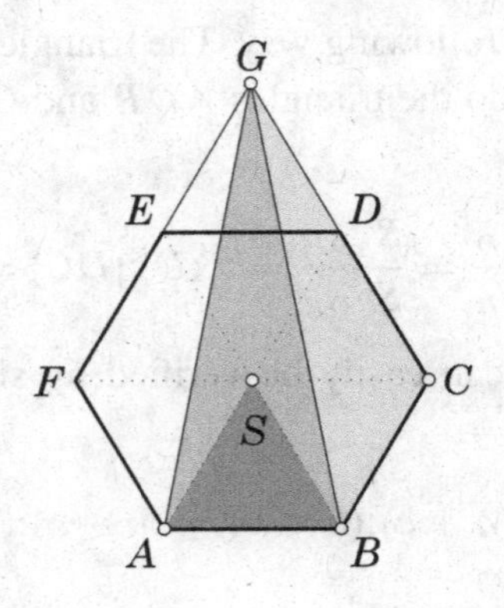

It therefore follows that the area of the pentagon $ABCGF$ is equal to $P + Q = P + \frac{1}{6}P = \frac{7}{6}P$.

We now let S denote the centre of the regular hexagon $ABCDEF$. The area R of the isosceles triangle ABG is three times that of ABS, because the perpendicular distance from G to the common side AB of the triangles ABG and ABS is three times that from S. We therefore have

$$R = 3Q = 3 \cdot \frac{P}{6} = \frac{P}{2}.$$

Now, in order to calculator the area T of the triangle BCG, we can apply the symmetry of the pentagon $ABCGF$ with respect to the perpendicular bisector of AB. From this, we immediately see that $2T + R = 7Q$ holds, which implies

$$T = \frac{7Q - R}{2} = \frac{1}{2}\left(\frac{7P}{6} - \frac{P}{2}\right) = \frac{P}{3}.$$

The triangles ABG and BCG thus have areas $\frac{1}{2}P$ and $\frac{1}{3}P$, respectively.

Remark. Note that the area T of BCG can also be computed independently in a similar way to the area of ABG, as the perpendicular distance from G to the side BC in BCG is twice the perpendicular distance from S to BC. □

5.6. Concurrency and Collinearity

Many problems in plane geometry concern themselves with proving that three lines pass through a common point, or alternately that three points lie on a common line, under certain specific circumstances. In this section, we present a few such problems from the Duel.

PROBLEMS

167. Let ABC be an acute-angled triangle. Let E be the perpendicular foot of A on BC and F the perpendicular foot of B on AC. Furthermore, let M be the midpoint of BE and N the midpoint of AF. Prove that the line through M perpendicular to AC and the line through N perpendicular to BC intersect in the midpoint of EF.

168. We are given a right triangle ABC. Its legs BC and AC are simultaneously the hypotenuses of two right-angled isosceles triangles BCP and ACQ erected outside ABC. Let D be the vertex of the right-angled isosceles triangle ABD with hypotenuse AB erected inside ABC. Prove that the point D lies on the line PQ.

SOLUTIONS

Problem 167. (B-T-2-14) *Let ABC be an acute-angled triangle. Let E be the perpendicular foot of A on BC and F the perpendicular foot of B on AC. Furthermore, let M be the midpoint of BE and N the midpoint of AF. Prove that the line through M perpendicular to AC and the line through N perpendicular to BC intersect in the midpoint of EF.*

Solution: Since M is the midpoint of the segment BE, the line perpendicular to AC (and thus parallel to BF) passing through M intersects EF in its midpoint G. Similarly, the line passing through N and perpendicular to BC (and thus parallel to AE) intersects EF in its midpoint G as well, which completes the proof.

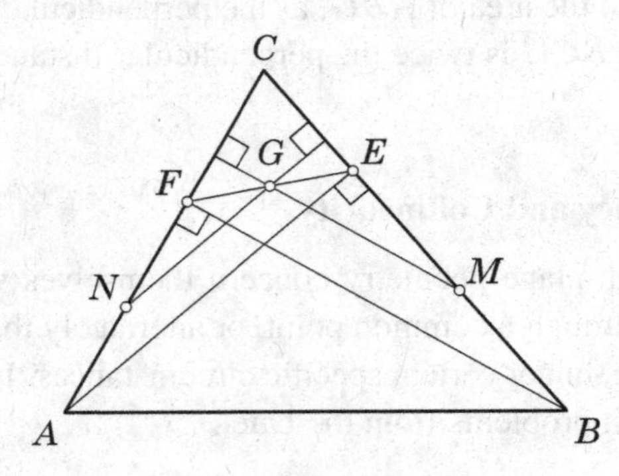

Another solution: The homothety with centre in E and ratio $\frac{1}{2}$ maps B to M and the altitude of the triangle ABC through B to the line perpendicular to AC through M. Since this homothety also maps F to the midpoint of EF, we see that the line through M perpendicular to AC passes through the midpoint of EF. By analogous reasoning with the centre of homothety in F, mapping the altitude in A to the line perpendicular to BC and passing through N, we see that the two perpendicular lines both pass through the midpoint of EF, as claimed. □

Problem 168. (B-T-2-16) *We are given a right triangle ABC. Its legs BC and AC are simultaneously the hypotenuses of two right-angled isosceles triangles BCP and ACQ erected outside ABC. Let D be the vertex of the right-angled isosceles triangle ABD with hypotenuse AB erected inside ABC. Prove that the point D lies on the line PQ.*

Solution: First of all, we can see that the vertex C of the given right-angled triangle lies on the line PQ, because $\angle QCA = \angle PCB = 45°$ implies

$$\angle QCP = \angle QCA + \angle ACB + \angle BCP = 45° + 90° + 45° = 180°.$$

If the given right-angled triangle ABC is isosceles, the statement of the problem is therefore obviously fulfilled.

Without loss of generality we can now assume $|AC| > |BC|$, as shown in the figure below. Let us consider the common point R ($R \neq C$) of PQ with the circle k with the centre S and the diameter AB. We shall prove that $R = D$.

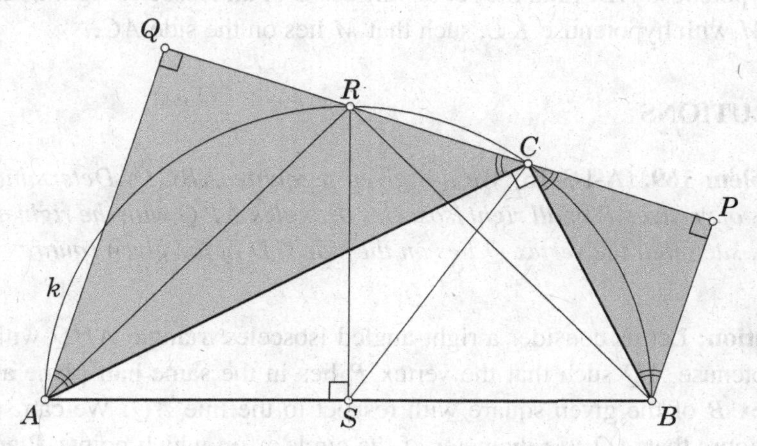

Since R lies on PQ, we certainly have

$$\angle RCA = \angle QCA = 45°.$$

Moreover, since R, C and A all lie on the same circle with centre S, we have

$$\angle RSA = 2 \cdot \angle RCA = 2 \cdot 45° = 90°.$$

This implies that the point R is the point on k which also lies on the perpendicular bisector of AB. ABR is therefore the right-angled isosceles triangle with hypotenuse AB, and thus $R = D$ holds, as claimed. □

5.7. Locus Problems

A type of problem that was far more popular in competitions a few decades ago concerns the definition of the set of points fulfilling a certain condition. This is still an interesting occasional subject in the Duel, as we see from the following problems.

PROBLEMS

169. We are given a square $ABCD$. Determine the locus of vertices P of all right isosceles triangles APQ with the right angle at P, such that the vertex Q lies on the side CD of the given square.

170. We are given an isosceles right triangle ABC. Let K be the midpoint of its hypotenuse AB. Find the set of vertices L of all isosceles right triangles KLM with hypotenuse KL, such that M lies on the side AC.

SOLUTIONS

Problem 169. (A-I-3-16) *We are given a square $ABCD$. Determine the locus of vertices P of all right isosceles triangles APQ with the right angle at P, such that the vertex Q lies on the side CD of the given square.*

Solution: Let us consider a right-angled isosceles triangle APQ with the hypotenuse AQ such that the vertex P lies in the same half-plane as the vertex B of the given square with respect to the line AQ. We can see in the figure that AQ is a diameter of the circle k, on which points P and D lie, since we have $\angle ADQ = \angle APQ = 90°$. In other words, A, P, Q and D lie (in this order) on the same circle, i.e., the quadrilateral $APQD$ is cyclic. This implies $\angle ADP = \angle AQP = 45°$, and P therefore lies on the diagonal BD of the square. B and S are the boundary positions of the points of P on the diagonal, with S resulting for $Q = D$ and B resulting for $Q = C$.

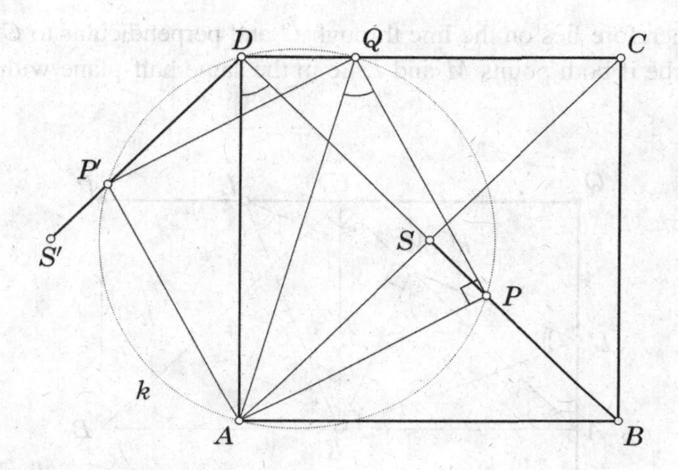

Conversely, it is clear that for every point P of the segment BS there exists a point Q of the side CD which is a vertex of a right-angled isosceles triangle APQ with the right angle at P.

Now, let us consider a right-angled isosceles triangle $AP'Q$ with the hypotenuse AQ, such that its vertex P' lies in the same half-plane as D with respect to the line AQ. As in the previous case, we can easily see that $AQDP'$ is a cyclic quadrilateral with $\angle ADP' = 45°$, and the locus of the vertices P' of all right-angled triangles resulting in this case is therefore the line segment DS', where S' is the point symmetric to the centre S of the given square with respect to AD.

In summary, we see that the locus of the vertices P of all right-angled isosceles triangles with a right angle at P is the pair of closed segments BS and DS'. □

Problem 170. (B-T-2-12) *We are given an isosceles right triangle ABC. Let K be the midpoint of its hypotenuse AB. Find the set of vertices L of all isosceles right triangles KLM with hypotenuse KL, such that M lies on the side AC.*

Solution: Let the points M and L be situated as in the picture below. Since $\angle KLM = \angle KCM = 45°$, the points K, L, C and M lie on a common circle k. We therefore have

$$\angle KCL = \angle KML = 90°,$$

and L therefore lies on the line through C and perpendicular to CK. This is also true if both points M and L lie in the same half-plane with respect to CK.

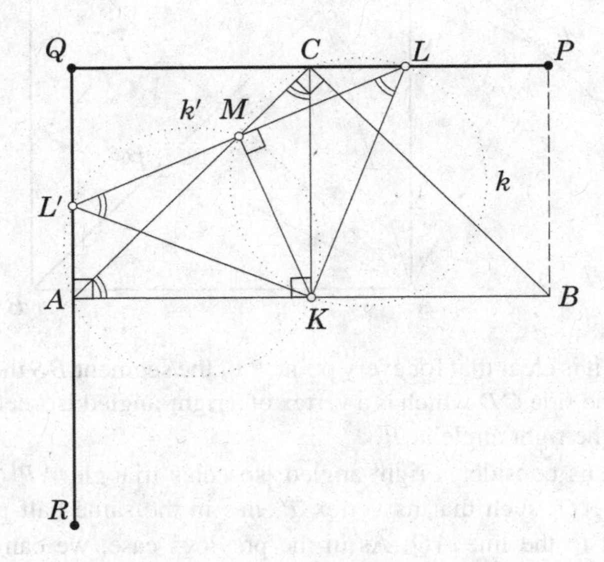

Now, let us consider the vertex L' in the opposite half-plane with respect to KM (see picture). Then the points K, M, L' and A similarly lie on a common circle k', with the diameter KL'. Moreover, the circles k and k' have equal diameters KL and KL'.

This implies that the point L must lie either on the segment PQ or on the segment QR.

Conversely, it is easy to see that for an arbitrary point L on either PQ or QR, there exists a unique point M on AC such that KLM is a right-angled isosceles triangle with hypotenuse KL. The set of all points L with the required property is therefore the union of the two line segments PQ and QR, with $PQ \perp QR$ and for which C and A are the midpoints of PQ and QR, respectively. □

5.8. Solid Geometry

Finally, there is the topic of solid geometry. In the early days of international olympiads, this was a topic that could be found on most every competition paper, but that, too, has changed. At the Mathematical Duel, this type of problem is still occasionally seen, however, and here we have some of the more interesting of these problems.

PROBLEMS

171. We are given a tetrahedron $ABCD$ with pairwise perpendicular edges at vertex D. Let K, L, M be the midpoints of edges AB, BC, CA, respectively. Prove that the sum of the measures of the angles at vertex D in three adjacent faces of the tetrahedron $KLMD$ is equal to $180°$.

172. Four unit spheres fit into the interior of a cylinder of minimal volume. Determine the dimensions and the volume of the cylinder.

SOLUTIONS

Problem 171. (A-I-2-14) *We are given a tetrahedron $ABCD$ with pairwise perpendicular edges at vertex D. Let K, L, M be the midpoints of edges AB, BC, CA, respectively. Prove that the sum of the measures of the angles at vertex D in three adjacent faces of the tetrahedron $KLMD$ is equal to* $180°$.

Solution: Since ABD, BCD and CAD are right triangles, we have

$$|KD| = \tfrac{1}{2}\,|AB|, \quad |LD| = \tfrac{1}{2}|BC|, \quad \text{and} \quad |MD| = \tfrac{1}{2}\,|CA|.$$

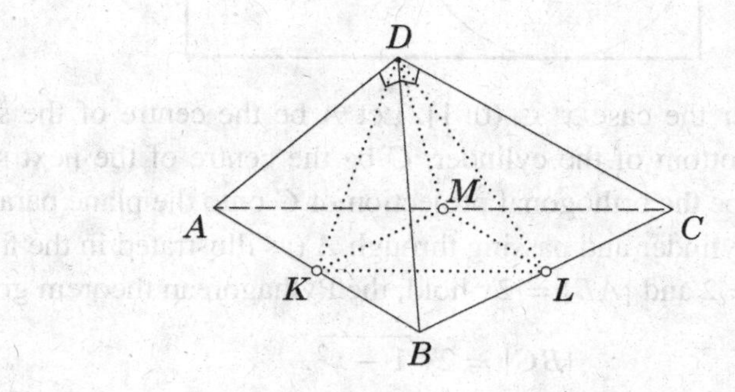

Since $|KD| = |KB|$ and $|LD| = |LB|$, we see that the triangles KLD and KLB are congruent. Similarly, the pairs of triangles LMD, LMC and

MKD, MKA are also congruent. We therefore have

$$\begin{aligned}\angle KDL + \angle LDM + \angle MDK &= \angle KBL + \angle LCM + \angle MAK \\ &= \angle ABC + \angle BCA + \angle CAB = 180^\circ,\end{aligned}$$

which completes the proof. □

Problem 172. (B-T-2-97) *Four unit spheres fit into the interior of a cylinder of minimal volume. Determine the dimensions and the volume of the cylinder.*

Solution: Let R and H denote the radius and the altitude of the cylinder, respectively. We shall prove that the cylinder with the minimal volume is the cylinder with the radius $R = 1$ and the altitude $H = 8$. The volume of this cylinder is equal to 8π.

Let $x > 0$ denote the distance of the centre of a sphere from the axis of the cylinder.

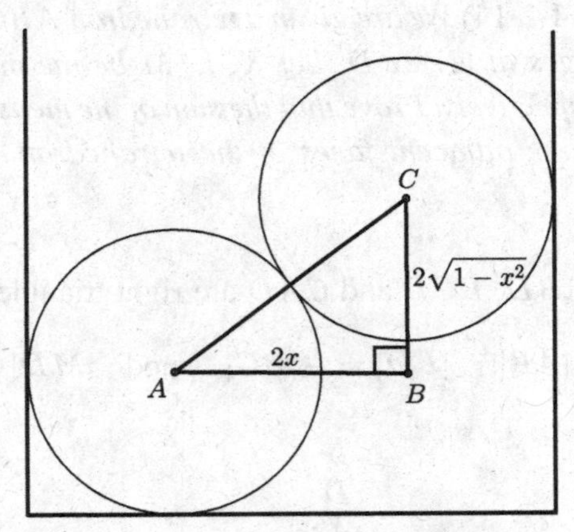

We first consider the case $x \in (0; 1]$. Let A be the centre of the sphere tangent to the bottom of the cylinder, C be the centre of the next sphere above it, and B be the orthogonal projection of C onto the plane parallel to the base of the cylinder and passing through A (as illustrated in the figure). Because $|AC| = 2$ and $|AB| = 2x$ hold, the Pythagorean theorem gives us

$$|BC| = 2\sqrt{1 - x^2}.$$

Since this difference in altitudes of the centres of successive centres of the spheres will always be the same, the altitude H of the cylinder containing the four unit spheres must fulfil the inequality

$$H \geq h = 2 + 3 \cdot |BC| = 2 + 6\sqrt{1 - x^2}.$$

Calculating the volume therefore gives us

$$V(x) = \pi(1 + x)^2 h \geq \pi(1 + x)^2 h = \pi(1 + x)^2 \cdot (2 + 6\sqrt{1 - x^2}).$$

Noting that

$$\sqrt{1 - x^2} \geq 1 - x \quad \text{and} \quad 3(1 + x)^2 > x + 3$$

must hold for $x \in (0; 1]$ therefore yields

$$\begin{aligned} V(x) - 8\pi &= 2\pi((1 + x)^2 + 3(1 + x)^2\sqrt{1 - x^2} - 4) \\ &= 2\pi(3(1 + x)^2\sqrt{1 - x^2} + (x - 1)(x + 3)) \\ &> 2\pi((x + 3)\sqrt{1 - x^2} + (x - 1)(x + 3)) \\ &= 2\pi(x + 3)(\sqrt{1 - x^2} + (x - 1)) \\ &\geq 0, \end{aligned}$$

and therefore $V(x) > 8\pi$.

In the case $x > 1$, we have $H \geq h = 2$ and therefore also obtain $V(x) > 8\pi$. In all possible cases, $V(x) > 8\pi$ therefore holds for $x > 0$. The cylinder with the minimal volume, which contains four unit spheres is therefore the cylinder with the radius $R = 1$ and the altitude $H = 8$, as claimed. □

Chapter 6

4! Years of Problems

This chapter contains all problems posed in the first $4! = 24$ years of the Mathematical Duel. Note that the standard structure of A, B and C categories was not introduced until the second year, and the first year therefore only consists of one set of questions. Also, it was common in early years for the same problem to be used in more than one category. This practice became more and more rare as the years went on.

Duel I: Bílovec, 1993

Individual

1. Prove that the number

$$1993^2 - 1992^2 + 1991^2 - 1990^2 + \ldots + 3^2 - 2^2 + 1^2$$

is divisible by 1993.

2. Let a, b, c be the lengths of the sides of a triangle with $a^8 + b^8 + c^8 = a^4b^4 + b^4c^4 + c^4a^4$. Prove that the triangle is equilateral.

3. Let $n \geq 6$ be an integer. Prove that a given square can be cut into n (not necessarily congruent) squares.

4. Let D be an interior point of the side AB of a triangle ABC and let E be the common point of the segment AD with the common external tangent of the incircles of the triangles ABD and ADC. Prove that the length of the segment AE is constant for a given triangle ABC and determine its length.

Team

5. A quadrilateral with the area Q is cut into four triangles with areas A, B, C and D by its diagonals. Prove that

$$A \cdot B \cdot C \cdot D = \frac{(A+B)^2 \cdot (B+C)^2 \cdot (C+D)^2 \cdot (D+A)^2}{Q^4}$$

holds.

6. Do 20 successive composite positive integers exist, none of which is greater than 20 000 000? If so, give an example of such numbers.

7. A triangle T is divided into three triangles T_1, T_2 and T_3, such that T is first cut into two parts along a median and one of these parts is then again cut along a median. Determine the lengths of the sides of the triangle T, if one of the triangles T_1, T_2 and T_3 is an equilateral triangle with sides of unit length.

Duel II: Chorzów, 1994

C-I

1. Determine all possible values of the expression $m+n+p+q$, where m, n, p, q are pairwise different positive integers fulfilling the equation

$$(7-m) \cdot (7-n) \cdot (7-p) \cdot (7-q) = 4.$$

2. Prove that the triple $(x, y, z) = (5, 11, 12)$ cannot be a solution of the equation

$$x^n + y^n = z^n$$

for any positive integer n.

3. In a circle k with radius r and the centre S, we are given a chord AB, which is not its diameter. On the ray opposite to BA, we are given a point C with $|BC| = r$. Let D be the common point of the segment CS and the circle k. Determine the angles $\alpha = \angle ADS$ and $\beta = \angle BCS$.

4. We are given a triangle ABC with $|AB| = 10$, $|AC| = 8$, $|BC| = 6$. On sides AC and BC of the triangle, we choose points K and L respectively, such that the circle passing through K, L, C is tangent to AB. What is the minimal distance between the points K and L?

C-T

1. Let a, b, c be the lengths of the sides of a triangle, fulfilling the condition

$$a^2 + b^2 + c^2 = ab + bc + ca$$

Prove that the triangle must be equilateral.

2. Show a manner in which it is possible to cut a given square into (a) 29, (b) 33, (c) 37 not necessarily congruent squares.

3. Let $n \geq 1$ be a positive integer, such that $2n + 1$ and $3n + 1$ are perfect squares. Decide whether it is possible for $5n + 3$ to be a prime number.

B-I

1. Determine the values of $x^{14} + x^{-14}$ and $x^{15} + x^{-15}$ if the equality $x^2 + x + 1 = 0$ holds.

2. Determine all pairs (x, y) of integers fulfilling the following system of inequalities:

$$x^2 + x + y \leq 3,$$
$$y^2 + y + x \leq 3.$$

3. Prove that the following inequality holds in every right-angled triangle with legs a, b and hypotenuse c:

$$a^3 + b^3 < c^3.$$

4. In the plane, we are given a straight line p and points A and B ($A \notin p$, $B \notin p$) on opposite sides of p. Construct two points X, Y on the line p, such that the triangle AXY is isosceles with base AY and the line BX is the angle bisector of AXY.

B-T

1. We wish to cut a cube into n separate small cubes using planes parallel to the faces. Determine the smallest number k, such that this is possible for all integers $n \geq k$.

2. Determine all pairs (x, y) of integers fulfilling the equation

$$\sqrt{x} + \sqrt{y} = \sqrt{90}.$$

3. Determine all positive integers n, such that

$$\frac{19n + 17}{7n + 11}$$

is an integer.

A-I

1. Three spheres with the same radius r are placed into a bowl in the form of half-sphere with the radius R. All of these spheres touch each other, and also touch a flat lid that is placed on the bowl. Determine the radius r of the small spheres in terms of R.

2. Determine all triples (a, b, c) of real numbers solving

$$a + bc = 2,$$
$$b + ca = 2,$$
$$c + ab = 2.$$

3. We are given an equilateral triangle with area S. Lines parallel to the sides of the triangle are drawn through a point M, which is given in the interior of the triangle. As shown in the figure, the areas of the resulting triangles are equal to S_1, S_2 and S_3. Determine the set all points M such that $S_1 + S_2 + S_3 = \frac{1}{2}S$ holds.

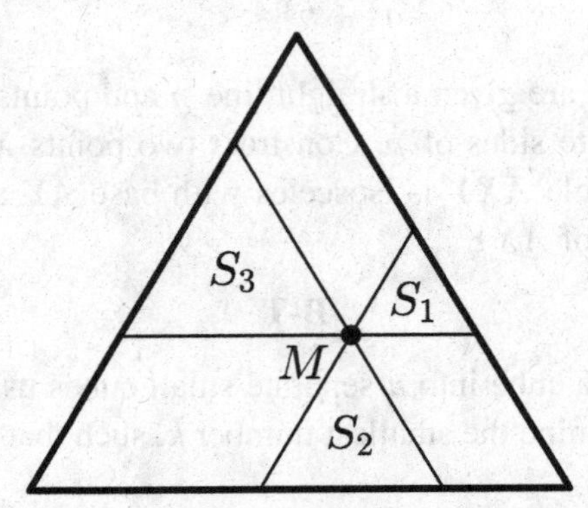

4. Determine $x^n + x^{-n}$ in terms of n and α, if we are given the relationship $x + \frac{1}{x} = 2 \cdot \cos\alpha$.

A-T

1. See B-T-1.

2. See B-T-2.

3. Let x, y, z be positive real numbers fulfilling the following inequalities:

$$\frac{1}{3} \leq xy + yz + zx \leq 3.$$

Determine which values can be achieved for the expressions

(a) xyz, and
(b) $x + y + z$.

Duel III: Bílovec, 1995

C-I

1. We bought 40 stamps with the face values of 1 crown, 4 crowns and 12 crowns. How many stamps of each kind did we buy, if we paid a total of 100 crowns?

2. We are given a right-angled triangle ABC with right angle in C. The circle with the diameter BC intersects the hypotenuse AB in a point D, such that $3|AD| = |DB|$ and $|CD| = 3$ cm hold. Determine the length of the median at the vertex A.

3. We place three spheres with the same radius into a (standing) cylinder with the given radius R. The spheres are mutually tangent and in contact with the base and lid of the cylinder, as well as the surface of the cylinder. How long is the longest straight stick (line segment) that can be placed in the cylinder along with the spheres?

4. Determine all positive integers divisible by 8, such that the sum of their decimal digits is equal to 7, and the product of their digits is equal to 6.

C-T

1. In a isosceles trapezoid $ABCD$, the longer base has the length of 4 cm. Its diagonal are perpendicular and their one is twice as long as the other. Determine

(a) the area of the trapezoid, and
(b) the radius of its circumcircle.

2. Prove that, for every integer number $n > 1$, the inequality

$$\left(1+\frac{1}{3}\right)\cdot\left(1+\frac{1}{8}\right)\cdot\left(1+\frac{1}{15}\right)\cdot\ldots\cdot\left(1+\frac{1}{n^2-1}\right)<2$$

holds.

3. On the side BC of an isosceles triangle ABC with $|AC| = |BC|$, we choose points M and N, with N closer to B than M, such that $|NM| = |AN|$ and $\angle MAC = \angle BAN$. Determine $\angle CAN$.

B-I

1. Determine all pairs (x, y) of real numbers fulfilling the equation

$$\left(x+\frac{1}{x}\right)\cdot(2+\sin y)=2.$$

2. Determine all positive integers m and n, such that $n+m^2$ and $m+n^2$ are both perfect squares.

3. Four spheres with a common radius r are placed in the interior of a regular tetrahedron with the edges of length a. The spheres are all tangent to each other and also to the faces on the tetrahedron. Determine r in terms of a.

4. Two circles O_1 and O_2 in the plane are externally tangent. They touch a common tangent in points A and D. AB is a diameter of O_1, and a line through B is tangent to O_2 in a point C. Prove that $|AB| = |BC|$ must hold.

B-T

1. Determine all non-constant polynomials $f(x)$ such that

$$f(x)\cdot\left(f(x^2)-x^2\right)=f(x^3)$$

holds for every $x \in \mathbb{R}$.

2. The diagonals AC and BD of the trapezoid $ABCD$ ($AB \parallel CD$) are perpendicular, and the length of the mid-segment of the trapezoid is equal to m. We choose a point M on the longer base AB such that $|AM| = m$ holds. Determine the length of the segment MC.

3. Determine as many triples of pairwise different positive integers (k, l, m) as possible, fulfilling the equation

$$\frac{1}{1995} = \frac{1}{k} + \frac{1}{l} + \frac{1}{m}.$$

A-I

1. Determine the value of the number a, if we are given

$$2 \cdot 4^x + 6^x = 9^x,$$

and $x = \log_{\frac{2}{3}} a$.

2. See B-I-2.

3. See B-I-3.

4. Let α, β, γ be the interior angles of a triangle. Prove that

$$\sin\alpha - \sin\beta + \sin\gamma = 4\sin\frac{\alpha}{2}\cos\frac{\beta}{2}\sin\frac{\gamma}{2}$$

holds.

A-T

1. The least common multiple of positive integers $5n - 1$ and $7n + 1$ is equal to $64n - 2$. Determine all possible values of $n \in N$.

2. Let $\alpha, \beta, \gamma, \delta$ be angles from the interval $\left[-\frac{\pi}{2}, \frac{\pi}{2}\right]$, such that

$$\sin\alpha + \sin\beta + \sin\gamma + \sin\delta = 1, \text{ and}$$

$$\cos 2\alpha + \cos 2\beta + \cos 2\gamma + \cos 2\delta \geq \tfrac{10}{3}$$

hold. Prove that $\alpha, \beta, \gamma, \delta \in [0, \frac{\pi}{6}]$ must hold.

3. We are given a triangle ABC, with sides of length a, b, c. Determine the point D in the plane of the triangle, such that the area of the convex quadrilateral $ABCD$ is twice as large area as the area of ABC and the circumference has the smallest possible value.

Duel IV: Chorzów, 1996

C-I

1. We consider three digit numbers that can be written using three different digits from the set $\{1, 2, \ldots, 9\}$. Prove that the sum all such numbers is divisible by 37.

2. We are given a square $ABCD$. Let points K, L, M, N be the midpoints of the sides AB, BC, CD, DA, respectively. Furthermore, let P, Q, R, S be the points of intersection of the lines AL and BM, BM and CN, CN and DK, DK and AL, respectively.

(a) Prove that $PQRS$ is a square.
(b) Determine the ratio of the area of $ABCD$ to the area of $PQRS$.

3. Determine all integer solutions of the equation

$$6x^2 + 5y^2 = 74.$$

4. Two congruent cubes are placed as in the figure below. Determine the angle $\angle XYZ$.

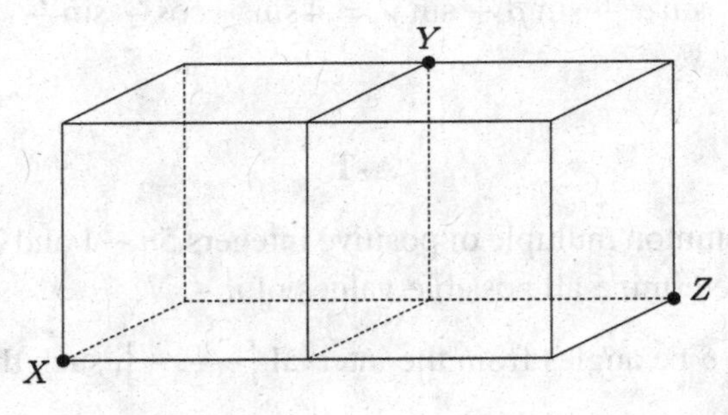

C-T

1. Prove that the inequality

$$2x^4 - 2x + 1 > 0$$

holds for every real number x.

2. The radius of the base of a cone is R and the slant height is equal to L. Determine the radius r of the sphere that can be inscribed in the cone.

3. Let a and b be positive integers such that

$$a + b + ab = 154$$

holds. Determine all possible values of the sum $a + b$. Explain your answer.

B-I

1. Let a, b, c be the lengths of the sides of a triangle. Prove the following inequality:

$$a^2 + b^2 + c^2 < 2(ab + bc + ca).$$

2. Every point of the plane is coloured in one of two colours. Prove that there exist among them two points of the same colour such that their distance is equal to 1996.

3. We are given the set of all seven-digit numbers, in which the digits are some permutation of 1, 2, 3, ..., 7. Prove that the sum of all these numbers is divisible by 9.

4. We are given a trapezoid $ABCD$ with perpendicular diagonals, as shown in the figure below. Determine all possible values of $pq + rs$, if the lengths of the bases are equal to 57 and 35.

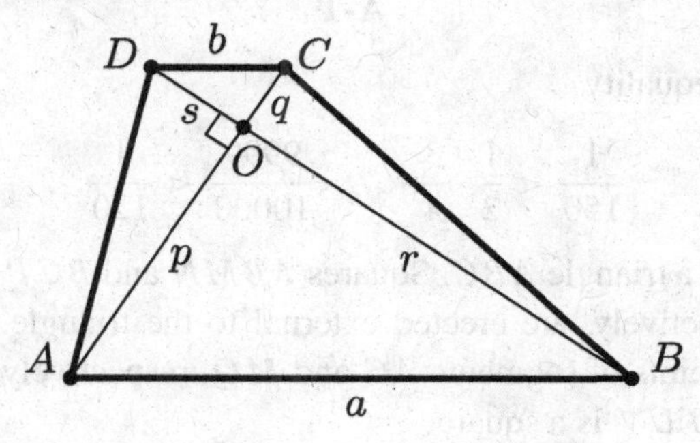

B-T

1. Determine all integer solutions of the equation

$$xyz + xy + yz + zx + x + y + z = 1996.$$

2. The radius of the base of a cone is R and the slant height is equal to L. A sphere of radius r_1 is inscribed in the cone and a sphere of radius r_2 is circumscribed about the cone. Determine the ratio $r_1 : r_2$.

3. Let p and q be a different positive integers. Prove that at least one of the equations

$$x^2 + px + q = 0 \quad \text{or} \quad x^2 + qx + p = 0$$

has a real root.

A-I

1. See B-I-2.

2. The ratio of the surface area of a cone to the surface area of its inscribed sphere is equal to 3:2. Determine the angle between a generatrix of the cone and the base of the cone.

3. Determine the values of the expression $x^2 + y^2 + z^2$, if it is known that the positive integers x, y, z fulfil the equations

$$7x^2 - 3y^2 + 4z^2 = 8 \quad \text{and} \quad 16x^2 - 7y^2 + 9z^2 = -3.$$

4. The medians of a triangle have the lengths 5 cm, $\sqrt{52}$ cm and $\sqrt{73}$ cm. Prove that the triangle is right angled.

A-T

1. Prove the inequality

$$\frac{1}{150} < \frac{1}{2} \cdot \frac{3}{4} \cdot \ldots \cdot \frac{9999}{10000} < \frac{1}{120}.$$

2. We are given a triangle ABC. Squares $ABMN$ and $BCPQ$, with centres T and V respectively, are erected external to the triangle. Let the S and U be the midpoints of segments AC and MQ, respectively. Prove that the quadrilateral $STUV$ is a square.

3. Three different real numbers a, b, c fulfil the condition

$$a + \frac{1}{b} = b + \frac{1}{c} = c + \frac{1}{a} = p,$$

where p is a real parameter.

(a) Determine which values are possible for p.
(b) Prove that $abc + p = 0$ must hold.

Duel V: Bílovec, 1997

C-I

1. Determine the last digit of the number

$$1 + 1 \cdot 2 + 1 \cdot 2 \cdot 3 + 1 \cdot 2 \cdot 3 \cdot 4 + \cdots + 1 \cdot 2 \ldots 1997.$$

2. In a trapezoid $ABCD$ with $AB \parallel CD$ and the area 225 cm^2, the diagonals AC and BD intersect in the point M and divide the trapezoid into four triangles. Determine the areas of these triangles, if we are given that the area of ABM is twice that of BCM.

3. Determine all real solutions of the following system of equations:

$$xy = z,$$
$$yz = x,$$
$$zx = y.$$

4. We are given a square $KLMN$. Prove the following equality for every point P lying on one of its diagonals:

$$|PK| \cdot |PM| + |PL| \cdot |PN| = |KL|^2.$$

C-T

1. Determine the smallest three-digit integer number n such that the sum of the digits of n is three times as large as the sum of the digits of the number $n + 69$.

2. Determine the ratio of the area of a regular pentagon inscribed in a given circle to the area of the regular pentagon circumscribed on the same circle.

3. The product of two polynomials $x^2 + ax + b$ and $x^2 + cx + d$ is the polynomial $x^4 + 4$. Determine the coefficients a, b, c, d.

B-I

1. Prove that the fraction

$$\frac{n^4 + 4n^2 + 3}{n^4 + 6n^2 + 8}$$

is not reducible for any integer n.

2. A quadrilateral $ABCD$ is inscribed in a circle. The diagonals AC and BD intersect at right angles in the point M and $|AC| = |BC|$ holds. Prove

$$|AB| \cdot |CD| = 2|AC| \cdot |MD|.$$

3. Prove, for an arbitrary pair of real numbers x, y, that $x^2 + y^2 \leq 2$ implies $|x| + |y| \leq 2$.

4. We are given an acute-angled triangle KLM with $|\angle KLM| < 60°$. Construct a polyline $KXYM$ with the smallest possible length, such that $X \in \overline{LM}$ and $Y \in \overline{KL}$.

B-T

1. Determine all triples of positive integers (x, y, z) fulfilling the equation

$$\frac{1}{x} + \frac{2}{y} + \frac{3}{z} = 4.$$

2. Four unit spheres fit into the interior of a cylinder of minimal volume. Determine the dimensions and the volume of the cylinder.

3. Solve the following system of equations in real numbers:

$$2x\left(1 + \frac{1}{x^2 + y^2}\right) = 1,$$

$$2y\left(1 - \frac{1}{x^2 + y^2}\right) = 3.$$

A-I

1. Prove the following inequality for any non-negative numbers (x, y, z):

$$z^2y(z - x) + y^2x(y - z) + x^2z(x - y) \geq 0.$$

When does equality hold?

2. We are given a right-angled triangle ABC. Let L denote the orthogonal projection of C onto the hypotenuse AB. We consider the half-circles k_1 and k_2 constructed externally on the legs AC and BC, respectively. Construct the line p passing through the vertex C and intersecting k_1 and k_2 in points K and M, respectively, such that the triangle KLM has the largest possible area.

3. Determine all functions $f : \mathbb{R} \to \mathbb{R}$ such that the following equation holds for every $x, y \in \mathbb{R}$:

$$f((x-y)^2) = (f(x))^2 - 2xf(y) + y^2.$$

4. Let $ABCD$ be a parallelogram, in which AC is the long diagonal. Let E and F be the orthogonal projections of the point C to the extensions of AB and AD, respectively. Prove that the following property holds:

$$|AB| \cdot |AE| + |AD| \cdot |AF| = |AC|^2.$$

A-T

1. Determine all triples (x, y, z) of positive integers: $x = (a)_{10}$, $y = (bc)_{10}$, $z = (def)_{10}$ with $xy = z$ and

$$\{a, b, c, d, e, f\} = \{2, 3, 4, 5, 6, 7\}.$$

2. Six unit spheres fit into the interior of a cylinder of minimal volume. Determine the dimensions and the volume of the cylinder.

3. We are given positive numbers a, b and c. Determine all real solutions of the following system of equations:

$$ax + by = (x-y)^2,$$

$$by + cz = (y-z)^2,$$

$$cz + ax = (z-x)^2.$$

Duel VI: Chorzów, 1998

C-I

1. Let a and b be real numbers, for which $a^2 + b^2 = 1$ holds. Determine the minimal value and the maximal value of the product $a \cdot b$. Explain your answer.

2. Let $A_1A_2A_3A_4$ be a trapezoid with $A_1A_2 \parallel A_3A_4$, and let P denote the point of intersection of the diagonals. Prove that

$$S_2 = \sqrt{S_1 \cdot S_3}$$

holds, where S_i denotes the area of the triangle $A_iA_{i+1}P$ $(i = 1, 2, 3)$.

3. Draw all non-congruent nets of the cube with edge-length 1 cm.

4. Determine whether it is possible to cover chessboards with the dimensions

(a) 11×11, and
(b) 12×12

completely, exclusively using shapes composed of three squares as shown in the figure below.

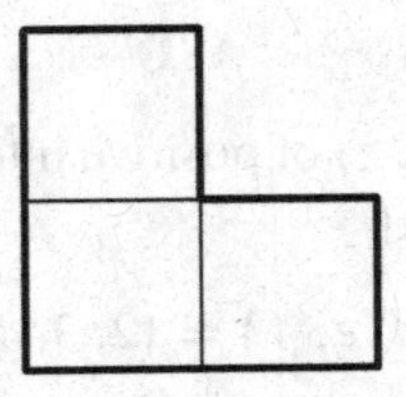

C-T

1. Prove that one of the nets of a tetrahedron is a triangle if and only if the opposite edges of the tetrahedron have the same length.

2. A bus ticket has a five-digit number from 00000 to 99999. We call the ticket *interesting*, if the sum of the first two digits is equal to:

(a) the sum of the last two digits,
(b) three times the sum of the last two digits,
(c) the sum of the squares of the last two digits,
(d) the sum of the cubes of the last two digits.

How many tickets are *interesting* in each case?

3. A cube, in which the interior is white and the faces of which are coloured red, is cut into n^3 small cubes of equal size. We call each resulting small cube *coloured*, if it has at least one red face. The number of *coloured* cubes is greater than the number of white ones. If n were increased by 1, the number of white cubes would be greater than the number of *coloured* ones. Determine the value of n.

B-I

1. Prove that the inequality

$$\frac{2}{a^2+b^2}+\frac{2}{b^2+c^2}+\frac{2}{c^2+a^2}\le\frac{1}{a^2}+\frac{1}{b^2}+\frac{1}{c^2}$$

holds for arbitrary real numbers a, b, c with $abc \neq 0$. When does equality hold?

2. The diagonals AC and BD of a trapezoid $ABCD$ with the bases AB and CD intersect in the point S. Determine the area of $ABCD$ if the areas of the triangles ABS and CDS are equal to $25\,\text{cm}^2$ and $16\,\text{cm}^2$, respectively.

3. We are given a right triangle ABC with right angle at the vertex C. Let P be an arbitrary interior point of the arc AB of the circumcircle of the triangle not containing the vertex C. Let X, Y, Z denote the orthogonal projections of P onto the lines AB, BC, CA, respectively. Prove that the points X, Y and Z lie on a straight line.

4. Prove that a two-digit number, which is not divisible by 10 but is divisible by the sum of its digits, is certainly divisible by 3. Is this also true for all three-digit numbers with this property?

B-T

1. Each vertex of a regular hexagon can be coloured either white or red. In how many different ways can the vertices of the hexagon be painted? (We consider two paintings as different if and only if there is no isometric mapping from one painting to the other.)

2. Let a, b, c, x, y, z be real numbers fulfilling the system of equations

$$\frac{x}{a}+\frac{y}{b}+\frac{z}{c}=1,$$
$$\frac{a}{x}+\frac{b}{y}+\frac{c}{z}=0.$$

Prove that this implies the following equality:

$$\frac{x^2}{a^2}+\frac{y^2}{b^2}+\frac{z^2}{c^2}=1.$$

3. Let $ABCD$ be a convex quadrilateral. Construct a quadrilateral $KLMN$ with the same area as the quadrilateral $ABCD$ such that the sides

KL, LM, MN, NK intersect the sides AB and BC, BC and CD, CD and DA, DA and AB respectively in points that divide every side of the quadrilateral $ABCD$ in the same ratio $x{:}y{:}x$.

A-I

1. Let a, b, c, d be a non-negative real numbers fulfilling the condition $a + b + c + d = 5$. Determine the maximum value of the expression $ab^2c^3d^4$.

2. Decide whether it is possible to place 50 circles in the plane such that each of them is externally tangent to at least $\frac{1}{7}$ of the others. Explain your answer.

3. We are given four spheres with the same radius r, each of which is tangent to the other three. These spheres are placed in the interior of a right circular cone such that its base is tangent to three of them and its surface is tangent to all four spheres. Determine the volume of the cone.

4. Determine all functions $f : \mathbb{R} \to \mathbb{R}$ such that the equation

$$f^2(x)f(y) = f(x - y)$$

holds for all $x, y \in \mathbb{R}$.

A-T

1. Determine all pairs (x, y) of integers fulfilling the equation

$$x^3 - 4x^2 - 5x - 6^y = 0.$$

2. We are given a trapezoid $ABCD$ with $AB \parallel CD$ such that $\angle ADB + \angle DBC = 180°$. Prove that this implies

$$|AB| \cdot |BC| = |CD| \cdot |DA|.$$

3. Let k be a non-negative integer. For which positive integers $n = 6k + 3$ is it possible to cover an $n \times n$ chessboard completely, using only tiles composed of three squares as shown in the figure below?

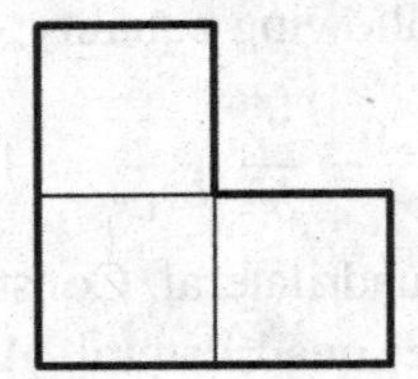

Duel VII: Graz, 1999

C-I

1. We are given that the sum of the squares of two positive integer is divisible by 3. Prove that each of the numbers must be divisible by 3.

2. We are given a triangle ABC. Let A' and B' be the midpoints of the sides BC and CA, respectively. If the points A, B, A' and B' lie on a common circle, prove that ABC is isosceles.

3. Some of the vertices of a cube are coloured in such a way that no four coloured vertices lie in a common plane. Determine the largest possible number of coloured vertices.

4. Determine the sum $a + b + c + d + e$, if we are given

$$1 \cdot 10^a + 2 \cdot 10^b + 3 \cdot 10^c + 4 \cdot 10^d + 5 \cdot 10^e = 315240.$$

C-T

1. Determine all quadruples (x, y, z, t) of integers, fulfilling the equation

$$\frac{1}{x^2} + \frac{1}{y^2} + \frac{1}{z^2} + \frac{1}{t^2} = 1,$$

and prove that no further solutions exist.

2. Let the diagonals AD and CG of a regular octagon $ABCDEFGH$ intersect at the point X. Prove that CDX is an isosceles triangle.

3. A clock is hanging on a wall. Only the points marking the full hours are visible on the face. The hour, minute and second hands of the clock are identical and move continuously (i.e., without jumps). A mirror is hung across from the clock, so that the reflection of the clock is visible. How many times a day does the clock appear in identical positions with its reflection?

B-I

1. We are given an acute-angled triangle ABC and a point P inside the triangle. Let A', B' and, C' be the points symmetric to P with respect to the sides BC, CA and AB, respectively. Prove that the hexagon $AC'BA'CB'$ is equilateral (all sides equal in length) if and only if P is the circumcentre of ABC.

2. Let us consider the system of the equations

$$x^2 + y^2 = a^{-1},$$

$$x \cdot y = a.$$

Determine the value of the real parameter a, if we know that the system of equations has exactly one solution in $\mathbb{R}^+ \times \mathbb{R}^+$. Determine this solution.

3. Prove that the inequality

$$(a^2 + b^2)^3 \geq (a^3 + b^3)^2$$

holds for any two real numbers a and b. When does equality hold?

4. Prove that the number $m^5 n - mn^5$ is divisible by 30 for all positive integers m and n.

B-T

1. A circle with radius r is inscribed in a rectangular trapezoid such that the larger base has the length $3r$. Calculate the area and the perimeter of the trapezoid.

2. Let x, y and z be real numbers such that $0 \leq x, y \leq z$. Prove that the following inequality holds:

$$xy(x + z) \leq (x + y)(yz + zx - xy).$$

When does equality hold?

3. Let X be a variable interior point of the segment AB. Two equilateral triangles AXC and XBD are erected in the same half-plane determined by the line AB. Determine the set of all points $E = AD \cap BC$.

A-I

1. Let V be the orthocentre of an acute-angled triangle ABC. We denote $a = |BC|$, $b = |AC|$, $c = |AB|$ and $x = |AV|$, $y = |BV|$, $z = |CV|$. Prove the following identity:

$$abc = ayz + bzx + cxy.$$

2. Let p be a positive integer number such that $2^p + 1$ is a prime number. Prove that a positive integer k exists such that $p = 2^k$ holds.

3. Prove that the inequality

$$a(a+b)+b(b+1) \geq -\tfrac{1}{3}$$

holds for all real numbers a and b. When does equality hold?

4. Determine all integer solutions of the equation

$$(x^2-y^2)^2 = 16y+1.$$

A-T

1. Determine all polynomials $P(x)$ such that

$$P^2(x) = P(xy)P\left(\frac{x}{y}\right)$$

holds for all real numbers $x, y \neq 0$.

2. In how many ways is it possible to choose three lattice points in a 4×4 square such that they are the vertices of a triangle? How many of these triangles are acute angled?

3. We are given a square $ABCD$ with the centre M. Let us consider circles k, which pass through the points A and M and intersect the segment MB in an interior point L other than M, the side AB in a point K other than A and the side DA in a point N other than A. Determine the largest possible area of the pentagon $AKLMN$.

Duel VIII: Bílovec, 2000

C-I

1. Determine all pairs (x, y) of positive integers x, y satisfying the following inequality:

$$1 \leq \frac{2}{x} + \frac{3}{y}.$$

2. In a right-angled triangle ABC with right angle in C and $\angle BAC > \angle ABC$, let L denote the point of intersection of the perpendicular bisector of AB with BC. Let K denote the point of intersection of the line AL and the altitude in C.

(a) Prove that the triangle KLC is isosceles.
(b) Under which circumstances is KLC equilateral?

3. Construct a triangle ABC with angles $\alpha = 45°$ and $\beta = 60°$, which is inscribed in a circle with radius 5 cm.

4. Let $m = \overline{ab}$ be a two-digit number and $n = \overline{ababab}$ a six-digit number. Prove that

(a) the number n is divisible by 3,
(b) the number n is divisible by 7,
(a) the number n is divisible by 9, if and only if m is divisible by 3.

C-T

1. The year 2000 has the property that

$$2000 = 16 \cdot 125 = 4^2 \cdot 5^3$$

is the product of a perfect square and a perfect cube, each of which is greater than 1. How many years from 1 to 1999 had this property?

2. We are given a line segment AB in the plane and a variable interior point Q. Equilateral triangles AQP and QBR are constructed in the same half-plane as determined by the line AB. Determine the set of all midpoints S of segments PR for all points Q.

3. Prove that the inequality

$$\frac{a^2 + b^2}{2} + \frac{b^2 + c^2}{2} \geq a\left(b - \frac{c}{2}\right) + c\left(b - \frac{a}{2}\right)$$

is satisfied for all real numbers a, b, c. When does equality hold?

B-I

1. Prove that no triple (x, y, z) of real positive numbers exists, satisfying the following system of inequalities:

$$|x - y| > z,$$
$$|y - z| > x,$$
$$|z - x| > y.$$

2. We are given two lines g_1 and g_2 that intersect in X and a point P not on either line. The point P_1 is symmetric to P with respect to g_1 and P_2 is symmetric to P with respect to g_2. Points Q_1 and Q_2 are symmetric to P_1 and P_2, respectively, with respect to X. We assume that Q_1, Q_2 and X are collinear. Prove that g_1 and g_2 are orthogonal.

3. Determine all possible values of (a, b, c, d) satisfying

$$a + 2b + c = 3,$$
$$b + 2c + d = 3,$$
$$c + 2d + a = 3,$$
$$d + 2a + b = 3.$$

4. A regular tetrahedron is inscribed in a sphere with radius R. A sphere with radius r is further inscribed in this tetrahedron. Determine the ratio $R{:}r$.

B-T

1. We are given a triangle ABC. Its interior angle BAC is divided into three congruent angles. The legs of these angles intersect the side BC in the points K and L. Determine the angle BAC, if four of the triangles with vertices in A, B, C, K and L are isosceles.

2. Solve the following equation in positive integers:

$$x^2 + y^2 = xy + 2x + 2y.$$

3. We are given two congruent circles k_1 and k_2 with a single common point. Construct a rectangle $ABCD$ such that $a{:}b = q$ holds for the lengths a and b of its sides and a given value $q > 0$. Both circles are to lie in the interior of the rectangle $ABCD$, and k_1 must be tangent to sides AB and AD, and k_2 must be tangent to sides CB and CD. For which values of q are both circles tangent to three sides of $ABCD$?

A-I

1. Determine all real numbers a for which the equations

$$x^4 + ax^2 + 1 = 0 \quad \text{and} \quad x^3 + ax + 1 = 0$$

have a common root.

2. A right triangle ABC with hypotenuse AB is given. Let K, L, M be the feet of the perpendiculars from an interior point E of ABC on the sides

BC, CA, AB, respectively. Determine a point E for which the area of the triangle KLM is maximal.

3. Prove that the inequality

$$\frac{ab}{cd}+\frac{bc}{da}+\frac{cd}{ab}+\frac{da}{bc}\geq\frac{a}{c}+\frac{b}{d}+\frac{c}{a}+\frac{d}{b}$$

is satisfied for all positive real numbers a, b, c, d. When does equality hold?

4. Let a, b, c be the lengths of the sides of a triangle. Determine the radii of the three excircles of the given triangle ABC.

A-T

1. We are given 1300 points inside a unit sphere. Prove that there exists a sphere with radius $\frac{2}{9}$ containing at least four of these points.

2. Solve the following equation in integers:

$$x^2+y^2=xy+4x+4y.$$

3. We are given two externally tangent circles $k_1(S_1;r_1)$ and $k_2(S_2;r_2)$. Construct a rectangle $ABCD$ with the following properties:

(a) The vertex D lies on the common inner tangent of k_1 and k_2.
(b) Both circles lie inside $ABCD$.
(c) The pairs of sides AB and AD, CB and CD are tangent to the circles k_1, k_2, respectively.

Determine the required conditions for a solution of this problem with respect to the values r_1 and r_2.

Duel IX: Chorzów, 2001

C-I

1. Let m be an integer. Determine all integer solutions of the equation

$$7(mx+3)=3(2mx+9).$$

Consider all possibilities for m.

2. The rectangle $ABCD$ with sides of the lengths a, b $(a>b)$ is given. Consider the rectangle $ACEF$ such that the vertex D lies on the segment EF.

Its sides CD and DA divide the rectangle $ACEF$ into three triangles whose areas are into the ratio 1:2:3. Determine the ratio a:b.

3. Determine all pairs (m, n) of positive integers such that the conditions

$$\langle m, n\rangle = 125 \quad \text{and} \quad [m, n] = 10\,000$$

are satisfied. (Here $\langle m, n\rangle$ and $[m, n]$ denote the greatest common divisor and least common multiple, respectively.)

4. The right triangle ABC with right angle at C is given. Let r be the inradius of ABC. Prove that the following relationship holds:

$$a + b = c + 2r.$$

C-T

1. Determine all triples (x, y, z) of prime numbers such that the following conditions hold:

$$x < y < z, \quad x + y + z = 77.$$

2. An A4 sheet of paper is a rectangle with area $\frac{1}{16}$ m^2 whose sides are in the ratio 1:$\sqrt{2}$. One such A4 sheet is placed on another such that they have a common diagonal, but are not identical. Determine the area of the resulting octagon.

3. Prove that the inequality

$$(a^3 + a^2 - a - 1)^2 - (a^3 - a^2 - a + 1)^2 \geq 0$$

is true for every non-negative real number a. When does equality hold?

B-I

1. Determine all integers x such that

$$f(x) = \frac{x^3 - 2x^2 - x + 6}{x^2 - 3}$$

is an integer.

2. We are given a convex quadrilateral in the plane. Prove that the lines passing through the midpoints of its opposite sides are perpendicular if and only if the diagonals of the given quadrilateral are of the same length.

3. Four real numbers a, b, c, d are given such that

$$b - a = c - b = d - c.$$

The sum of all these numbers is 2, and the sum of their cubes is $\frac{4}{3}$. Determine the numbers.

4. Determine how many triples (x, y, z) of positive integers satisfy the following system of inequalities:

$$x^2 + y^2 + z^2 \leq 2001,$$
$$\frac{x^2}{x^2 + 2yz} + \frac{y^2}{y^2 + 2zx} + \frac{z^2}{z^2 + 2xy} \leq 1.$$

B-T

1. Solve the following system of equations in real numbers:

$$x^2y + y^2z = 254,$$
$$y^2z + z^2x = 264,$$
$$z^2x + x^2y = 6.$$

2. The square $ABCD$ with sides of length 4 cm is given. Let K and L be the midpoints of its sides BC and CD, respectively. Determine the area of the quadrilateral whose sides lie on the lines AK, AC, BL and BD.

3. A quadratic equation $ax^2 + bx + c = 0$ has no real roots and we know that $a + b + c < 0$ holds. Determine the sign of the coefficient c.

A-I

1. Prove that

$$L = m^{21}n^3 - m^7n - m^3n^{21} + mn^7$$

is divisible by 42 for all integer values of m, n.

2. Find all functions $f : \mathbb{R} \to \mathbb{R}$ such that the equation

$$f(f(x + y)) = f(x) + y$$

is satisfied for arbitrary $x, y \in \mathbb{R}$.

3. Construct a quadrilateral inscribed in the circle with the radius 5 cm with opposite sides of the lengths 4 cm and 6 cm and with maximum area.

4. Let a, b, c be the lengths of the sides of a triangle. Prove that the following inequality holds:

$$3a^2 + 2bc > 2ab + 2ac.$$

A-T

1. Prove that the inequality

$$\frac{1}{ab(a+b)} + \frac{1}{bc(b+c)} + \frac{1}{ca(c+a)} \geq \frac{9}{2(a^3+b^3+c^3)}$$

is satisfied for any positive real numbers a, b, c. When does equality hold?

2. Let ABC be a triangle. Equilateral triangles ABD, BCE and CAF are erected external to ABC. Prove that the midpoints of these three triangles are the vertices of an equilateral triangle.

3. Is it possible for three different positive integers x, y, z to exist between two successive perfect squares, such that one is the geometric mean of the other two? In other words, is it possible that $n^2 < a, b, c < (n+1)^2$ and $c = \sqrt{ab}$ hold with a, b, c all different? If so, give an example. If not, prove why this is not possible.

Duel X: Graz, 2002

C-I

1. A big piece of chocolate is made of less than 200 but more than 150 small squares. If the chocolate is shared equally among seven children, one square remains. If it is shared equally among eight teachers, five squares remain. How many squares is the big piece of chocolate made of?

2. The equilateral triangle CDE has one side in common with a square $ABCD$. E lies outside the square and the sides of the square and the triangle are of length a. Determine the radius of the circumcircle of triangle ABE and prove that this is a correct value.

3. We are given a right-angled triangle ABC with right angle in C. Squares $ACDE$ and $BCFG$ are erected on the outside of ABC. Prove that $BFDA$

is an isosceles trapezoid (i.e., that two of its sides are parallel, and that the other two sides are of equal length).

4. We are given the following array of numbers:

1	2	3
2	3	1
3	1	2

In each move we are allowed to add 1 to all three numbers in any row or to subtract 1 from each of the three numbers in any column. Which of the following arrays can be the result of a number of such moves? If it is possible to reach this array, determine the moves that must be made. If it is not possible, explain why not.

2	2	5
1	1	1
4	2	2

2	2	3
2	2	1
4	1	3

C-T

1. Find the radius of the sphere passing through all six vertices of a regular triangular prism, the edges of which are all of equal length a.

2. We are given a bag full of T-tetrominoes, as shown in the figure below.

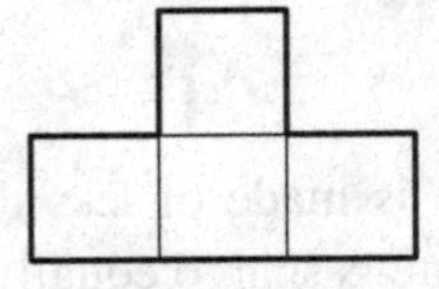

Is it possible to completely cover an 8×12 chessboard with T-tetrominoes without overlapping (assuming that the squares of the chessboard are the same size as the squares of the tetromino)? Is it possible for a 3×8 chessboard? Is it possible for a 7×10 chessboard? In each case, if it is possible, draw such a covering. If not, explain why it is not possible.

3. Adding three two-digit numbers yields a three-digit sum, as shown:

$$\overline{ab} + \overline{cd} + \overline{ef} = \overline{ghk}.$$

Different letters stand for different digits. What is the largest possible value of $\overline{ghk}$? Explain why your value is the largest possible one.

B-I

1. Find all real solutions of the following system of inequalities:

$$\frac{1}{2} \cdot x^2 + \frac{8}{y^2} \leq 4,$$
$$\frac{8}{x^2} + \frac{1}{2} \cdot y^2 \leq 4.$$

2. Let ABC be a right-angled triangle with right angle in C. Further, let M be the midpoint of the hypotenuse AB and F be the foot of the altitude through C on AB. Prove that the angle bisector of $\angle ACB$ and the perpendicular bisector of the shortest side of ABC intersect in the incentre of triangle CFM.

3. Find all real roots of the equation

$$2x^5 + x^4 + 4x^3 + 2x^2 + 2x + 1 = 0.$$

4. The floor of a room with area 5 m^2 is (partly or completely) covered using nine carpets, each of which has an area of exactly 1 m^2, but the shapes of which are arbitrary. Prove that there must exist two carpets that cover a common section of floor with an area of at least $\frac{1}{9}$ m^2.

B-T

1. Let n be a positive integer with $n \geq 4$. Prove that any given triangle can be cut into n isosceles triangles.

2. Let the coefficients of the quadratic equations

$$x^2 + p_1 x + q_1 = 0 \quad \text{and} \quad x^2 + p_2 x + q_2 = 0$$

fulfil the condition

$$p_1 p_2 = 2 \cdot (q_1 + q_2).$$

Prove that at least one of these equations has real roots.

3. We are given a right-angled triangle ABC with right angle in C. Let Q be the midpoint of AB and let P be a point on AC and R be a point on

BC such that $\angle PQR = 90°$. Prove that the triangles ABC and PQR are similar.

A-I

1. Let the equality

$$a^2 + b^2 + c^2 + d^2 + ad - cd = 1$$

be fulfilled for some real numbers a, b, c and d. Prove that $ab + bc$ cannot be equal to 1.

2. Prove that the expression $2^{12n+8} - 3^{6n+2}$ is divisible by 13 for every non-negative integer value n.

3. Let a and b be positive real numbers such that $a^2 + b^2 = 1$ holds. Prove

$$\frac{a}{b^2+1} + \frac{b}{a^2+1} \geq \frac{2}{3} \cdot (a+b).$$

When does equality hold?

4. See B-I-4.

A-T

1. Evaluate the following sums in terms of n:

$$\text{(a)} \quad \sum_{k=1}^{n} k! \cdot (k^2 + 1),$$

$$\text{(b)} \quad \sum_{k=1}^{n} 2^{n-k} \cdot k \cdot (k+1)!.$$

2. Let P be a point in the interior of the right-angled isosceles triangle ABC with hypotenuse BC such that the distances of P to the vertices of the triangle are given by

$$|AP| = a, \quad |BP| = a + b \quad \text{and} \quad |CP| = \sqrt{a^2 + b^2}$$

for given positive real numbers a and b. Prove that AP and BP are perpendicular.

3. Find all four-digit numbers n with the property that $2n$ is written in the same form (with the same digits in the same order) in some number system with base z as n is in base 10.

Duel XI: Bílovec, 2003

C-I

1. Determine the smallest positive integer n such that the remainders of n after division by 2, 3, 4, 5 and 6 are all equal to 1 and the remainder after division by 7 is equal to 4.

2. We are given 385 different points in the interior of a $3\,\text{m} \times 2\,\text{m}$ rectangle. Prove that there exist five points among these which can be covered by a square with sides of length 25 cm.

3. We are given real numbers a, b, for which $a + b = 1$ holds. Prove that the inequality

$$a^3 + b^3 \geq ab$$

is satisfied. When does equality hold?

4. We are given an isosceles triangle with sides of length 6 cm, 6 cm and 4 cm. Determine the distance between its circumcentre and its incentre.

C-T

1. Determine the maximum and the minimum among all sums

$$\overline{abc} = \overline{def} + \overline{ghi},$$

in which all of the digits $a, b, c, d, e, f, g, h, i$ are different.

2. We are given a pentagon $ABCDE$ with

$$|AC| = |AD| = 1, \quad |BC| = |DE| = \tfrac{1}{2},$$

$$\angle ABC = \angle AED = 90^\circ \quad \text{and} \quad |CD| = a.$$

(a) For which values of a is a pentagon $ABCDE$ convex (i.e., each interior angle less than 180°)?

(b) Determine the area of $ABCDE$ for all such values of a.

3. Determine all pairs (x, y) of positive integers such that the value of

$$\frac{2x + 2y}{xy}$$

is a positive integer.

B-I

1. Solve the following system of equations in real numbers:

$$1 + x + y = xy,$$
$$2 + y + z = yz,$$
$$5 + z + x = zx.$$

2. Let $ABCD$ be a parallelogram and M be a point on the line segment BC such that $|CM| = 1$ and $|BM| = k$ (where k is a positive real number). Furthermore, let N be the common point of AC and DM. Determine the value of k if

$$P_{ABMN} = \tfrac{1}{3} P_{ABCD}$$

holds. (P_{XYZU} denotes the area of the quadrilateral $XYZU$.)

3. The function $f(x) = ax^2 + bx + c$ has the property that

$$|f(-2)| = |f(0)| = |f(2)| = 2$$

holds. Determine all possible positive values of a, b and c.

4. We consider a line segment AB of the length c and all right triangles with hypotenuse AB. For all such right triangles, determine the maximum diameter of a circle with the centre on AB which is tangent to the other two sides of the triangle.

B-T

1. Determine all three-digit numbers $\overline{xyz}$ which are divisible by all three digits x, y, z, with no two digits being equal.

2. We are given a convex pentagon $ABCDE$, whose diagonals are each parallel to a side. We know that $|CE| = t \cdot |AB|$ for a positive real number t. Express

$$\frac{|AE|^2 - |BC|^2}{|BE|^2 - |AC|^2}$$

in terms of t if $|BE|^2 \neq |AC|^2$ holds.

3. Determine all positive integers n $(n \leq 200)$ such that there exist exactly six integers x for which

$$\frac{n+x}{2-x}$$

is an integer.

A-I

1. Solve the following system of equations in real numbers:

$$\sqrt{x^2+(y+1)^2}+\sqrt{(x-12)^2+(4-y)^2}=13,$$
$$5x^2-12xy=24.$$

2. Let CS be the median and CD the altitude in a triangle ABC, which is not isosceles. Prove that the triangle ABC is right angled if $\angle ACS = \angle DCB$ holds.

3. Prove that the polynomial

$$P_n(x)=3(2x-1)^{2n+2}-3x^{2n+2}-9x^2+12x-3$$

is divisible by the polynomial

$$Q(x)=3x^3-4x^2+x$$

for all positive integers n.

4. Determine the radius of the sphere inscribed in a tetrahedron with five edges of length 3 and one edge of length 4.

A-T

1. In the coordinate plane, determine the set of all points with coordinates (x, y), such that the real numbers x and y satisfy the inequality

$$\sin x \geq \cos y,$$

with $x, y \in [-4, 4]$.

2. Let $k_1(S_1; r_1)$ and $k_2(S_2; r_2)$ be two circles externally tangent at P. Furthermore, let XPY be a triangle and k be its circumcircle with centre S such that $X \in k_1$ and $Y \in k_2$. Prove that the line XY is a common tangent of both circles k_1 and k_2, if and only if, the following conditions are simultaneously satisfied:

(a) $\angle S_1SS_2 = 90°$, and
(b) k is tangent to the line S_1S_2.

3. Determine all positive integers n ($n \leq 50$) such that there exist eight integers x for which

$$\frac{x^2 + 7x - n}{x + 2}$$

is an integer.

Duel XII: Chorzów, 2004

C-I

1. We are given four congruent circles with radius r and centres in the vertices of a given square such that each neighbouring pair of circles is externally tangent. Determine the radius R of the circle, which has an internal point of tangency with each of the four smaller circles.

2. Find all real numbers a such that the root of the equation

$$a(x - 2) + x - 5 = 0$$

also satisfies the equation

$$2(a + 1)x + 5a + 8 = 0.$$

3. Let P be an arbitrary interior point of the diagonal AC of a rectangle $ABCD$. Furthermore, let K, L, M, N be the points of intersection of the lines through P and parallel to the sides of the rectangle with the sides AB, BC, CD, DA, respectively. Prove that

$$|LM| + |NK| = |AC|$$

holds.

4. Determine all pairs (x, y) of integers for which

$$x^2 + y^2 + 2x + 2y - 6 = 0$$

holds.

C-T

1. Determine the number of three-digit numbers m with the property that $m - m'$ is a positive integer divisible by 7. (The number m' is the three-digit number with the same digits as m written in inverse order.)

2. Let P be the midpoint of the side CD of a rectangle $ABCD$ and T the point of intersection of the lines AP and BD. Determine the ratio $|AB|:|BC|$ if $\angle ATB = 90°$.

3. Determine all pairs (x, y) of integers, solving the following system of inequalities:

$$x^2 + x + y \leq 2,$$

$$y^2 + y + x \leq 4.$$

B-I

1. Prove that for any rational number x, there exists a rational number y such that the equation

$$2x^3 - x^2y - 2xy^2 + y^3 + 3x^2 - 3y^2 - 2x + y - 3 = 0$$

holds.

2. In a parallelogram $ABCD$, P, Q, R and S are the midpoints of AB, BC, CD, DA, respectively. T is the midpoint of RS and X and Y are the points of intersection of QT with the lines PR and PC, respectively. Prove that the area of the quadrilateral $CRXY$ is equal to the sum of the areas of the triangles PXT and PQY.

3. Calculate the value of the sum

$$\left(\frac{1+2}{3} + \frac{4+5}{6} + \cdots + \frac{2002+2003}{2004}\right) + \left(1 + \frac{1}{2} + \frac{1}{3} + \cdots + \frac{1}{668}\right).$$

4. Solve the following system of equations in real numbers:

$$a^7 + a = 2b^4,$$

$$b^7 + b = 2c^4,$$

$$c^7 + c = 2a^4.$$

B-T

1. All faces of a white cube are coloured black. The cube is cut into n^3 small cubes using planes parallel to the faces of the cube. The number of all small white cubes is then less than the number of all coloured cubes (i.e., cubes with at least one black face). If we increase n by 1, the number of white small cubes will be greater than the number of all coloured cubes. Determine all possible values of n.

2. Solve the following system of linear equations with the real parameter a in real numbers:

$$x - a^3y + a^2z = -a^3x + y - az = a^2x - ay + z = 1.$$

3. Let ABC be an isosceles triangle with $|AC| = |BC|$. We are given a circle k with $C \in k$ and AB tangent to k. Furthermore, X is the point of tangency of AB and k, and P and Q are the points of intersection of AC and BC with k, respectively. Prove the following property:

$$|BQ| \cdot |AX|^2 = |AP| \cdot |BX|^2.$$

A-I

1. Let a, b, c be arbitrary positive real numbers. Prove that the inequality

$$\frac{1}{b(a+b)} + \frac{1}{c(b+c)} + \frac{1}{a(c+a)} \geq \frac{1}{2(a^2+b^2+c^2)}$$

holds. When does equality hold?

2. We are given the numbers $(15)_p$ and $(26)_p$ in some base $p \geq 2$, (p is a positive integer). Determine all values of p, such that the numbers $(15)_p$ and $(26)_p$ are relatively prime (i.e., have no common factor greater than 1).

3. Let ABC be an acute-angled triangle. Let D be the foot of the altitude from C on the side AB. Furthermore, let X and Y be the orthogonal projections of the point D on the sides BC and AC, respectively, and Z the common point of the line XY and the line parallel to BC through D. Prove that AZ is perpendicular to BC.

4. Determine all pairs (x, y) of positive integers such that

$$17x^2 + 1 = 9y! + 2004$$

A-T

1. Determine all functions $f : \mathbb{R} \to \mathbb{R}$ such that

$$f(xf(y) + x) = xy + f(x)$$

holds for all $x, y \in \mathbb{R}$.

2. We are given a quadrilateral $ABCD$ with AB parallel to CD and

$$|AB| = |AC| = |AD| = \sqrt{3} \quad \text{and} \quad |BC| = \sqrt{2}.$$

Determine the length of the diagonal BD.

3. A number is called *bumpy* if its digits alternately rise and fall from left to right. For instance, the number 36180 is bumpy, because $3 < 6$, $6 > 1$, $1 < 8$ and $8 > 0$ all hold. On the other hand, neither 3451 nor 81818 are bumpy.

(a) Determine the difference between the largest and smallest five-digit bumpy numbers.
(b) How many five-digit bumpy numbers have the middle digit 5?
(c) Determine the total number of five-digit bumpy numbers.

Duel XIII: Graz, 2005

C-I

1. How many right-angled triangles exist, whose sides have integer length and whose perimeter have the length 2005?

2. Prove that

$$x^2 + y^2 + 2 \geq (x+1)(y+1)$$

is true for all real values of x and y. For which values of x and y does equality hold?

3. We are given a triangle ABC, with $\alpha = 30°$ and $\beta = 60°$. Let S be the midpoint of AB, S_1 and S_2 the circumcentres of ASC and BSC, respectively, and r_1 and r_2 the respective radii of these circles.

(a) Prove that the triangles ABC and S_1S_2S are similar.
(b) Determine the ratio $r_1{:}r_2$.

4. The digits 1–9 are placed in a triangular array as shown such that the sums of the four numbers on each side are equal. The numbers 1, 2 and 3 are placed in the corners. Where can the number 9 not be placed?

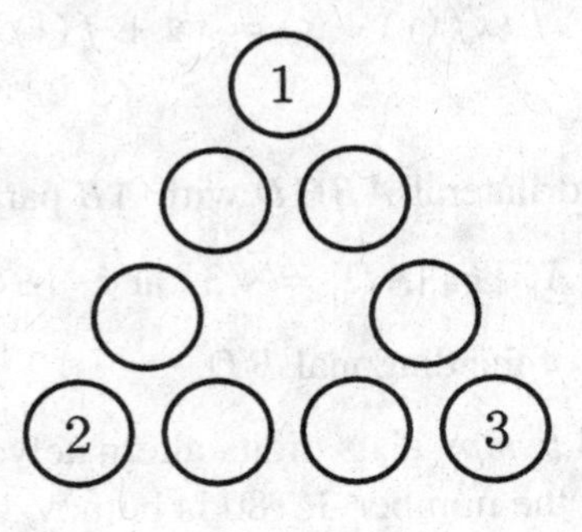

C-T

1. We are given a regular octagon $ABCDEFGH$. Drawing the diagonals AD, BE, CF, DG, EH, FA, GB and HC yields another smaller regular octagon in the interior of the original octagon. Determine the ratio of the areas of the original octagon and the resulting octagon.

2. Determine all positive integers z, for which positive integers m and n exist such that

$$z = \frac{mn+1}{m+n}.$$

3. Let H be the orthocentre of an acute-angled triangle ABC. The altitude in A has the length 15 cm, and this altitude divides BC into two sections of lengths 10 cm and 6 cm respectively. Determine the distance of H from the side BC.

B-I

1. Solve the following system of equations in real numbers:

$$x + \frac{1}{x} = 2y^2,$$

$$y + \frac{1}{y} = 2z^2,$$

$$z + \frac{1}{z} = 2x^2.$$

2. We are given a square $ABCD$. Construct points K, L, M and N on the sides AB, BC, CD and DA respectively, such that the pentagon $KLCMN$ has maximum area with KL and MN parallel to AC and NK parallel to BD.

3. Determine all right-angled triangles with perimeter 180 such that the lengths of the sides are integers.

4. We are given a triangle ABC with $\angle ABC = \beta$ and incentre I. Points A', B' and C' are symmetric to I with respect to BC, CA and AB, respectively. Prove that $\angle A'B'C' = \beta'$ is independent of the value of $\angle BAC$ and express β' in terms of β.

B-T

1. (a) A number x can be written using only the digit a both in base 8 and in base 16, i.e.,

$$x = (aa \ldots a)_8 = (aa \ldots a)_{16}.$$

Determine all possible values of x.

(b) Determine as many numbers x as possible that can be written in the form $x = (11 \ldots 1)_b$ in at least two different number systems with bases b_1 and b_2.

2. We are given an acute-angled triangle ABC with orthocentre O. Let F be the midpoint of AB and N be the point symmetric to O with respect to F. Show that $\angle ACO = \angle BCN$ must hold.

3. Determine an integer $a < 10000$ such that $a, a+1, a+2, \ldots, a+12$ are all composite numbers.

A-I

1. Let P be an arbitrary point in the interior of a right-angled isosceles triangle ABC with hypotenuse AB. Prove that a triangle exists with sides of lengths AP, BP and $CP\sqrt{2}$.

2. Let a, b, c be non-zero real numbers. Prove that the inequality

$$\frac{a^2}{b^2} + \frac{b^2}{c^2} + \frac{c^2}{a^2} \geq \frac{a}{b} + \frac{b}{c} + \frac{c}{a}$$

holds. For which values does equality hold?

3. Determine all integer solution of the following system of equations:

$$x^2z + y^2z + 4xy = 40,$$
$$x^2 + y^2 + xyz = 20.$$

4. Let ABC be an acute-angled triangle. Let D be the midpoint of BC and O the incentre of ABC. The line DO intersects the altitude in A in the point G. Prove that the length of AG is equal to the inradius of ABC.

A-T

1. We are given a square $ABCD$ with sides of length a and a line segment of length d with $a \leq d \leq a\sqrt{2}$. Construct a quadrilateral $KLMN$ with minimal perimeter such that the points K, L, M and N lie on the sides AB, BC, CD and DA, respectively and one diagonal of $KLMN$ has the length d.

2. Let $2s$ be the length of the perimeter of a triangle ABC and let ρ, r_a, r_b and r_c be the radii of the incircle and three excircles of ABC, respectively. Prove that the following inequality holds:

$$\sqrt{\rho \cdot r_a} + \sqrt{\rho \cdot r_b} + \sqrt{\rho \cdot r_c} \leq s.$$

3. Let V be a real function defined by the expression

$$V(x) = (x-1)(x-2) + (x-1)(x-2)(x-3)(x-4) + (x-3)(x-4).$$

(a) Determine the minimum value of $V(x)$.
(b) Determine all values of x for which this minimum value is assumed.

Duel XIV: Bílovec, 2006

C-I

1. Determine all positive integers n such that

$$n^3 - n + 3$$

is a prime number.

2. We are given a regular octagon $ABCDEFGH$. Determine with proof, which of the two triangles ADF and CEG has the larger area.

3. Let ABC be a right triangle with the hypotenuse AB of length c. Determine the sum $t_a + t_b + t_c$ of the lengths of its medians in terms of c.

4. Determine all pairs (x, y) of real numbers for which the inequality

$$4xy + 2y - x^2 - 5y^2 - 1 \geq 0$$

holds.

C-T

1. Determine all integer solutions $x \geq y \geq z$ of the equation

$$xyz = 16.$$

2. A square $ABCD$ of the size 7×7 is divided into 49 smaller congruent squares using line segments parallel to its sides. Determine the total number of paths from the vertex A to the vertex C along lines of the resulting grid, if we are only allowed to move in the directions of the vectors $\overrightarrow{AB}$ and $\overrightarrow{AD}$.

3. We are given a non-isosceles trapezoid $KLMN$ with the larger base KL. Prove the following statement: If the circumcircles of KLM and KLN are congruent, then their centres are symmetric with respect to the perpendicular bisector of KL.

B-I

1. Determine all integers $a > b > c > d > e$ such that

$$(5-a)(5-b)(5-c)(5-d)(5-e) = 20$$

holds.

2. Determine all right-angled triangles with integer side lengths, whose circumference is numerically equal to its area.

3. Determine all positive integers n such that

$$n^8 + n^4 + 1$$

is a prime number.

4. We are given an isosceles triangle ABC with the base AB. Let M be the foot of the altitude from the vertex A on its side BC. We are given that the equation $|CM| = 3|BM|$ holds. Prove that we can construct a right triangle from two line segments AB and the line segment AC.

B-T

1. Determine all pairs (a, b) of positive integers such that

$$\frac{2}{a^2} + \frac{3}{ab} + \frac{2}{b^2}$$

is an integer.

2. One large cube $ABCDEFGH$ is formed from eight small congruent cubes. Determine the total number of paths from the vertex A to the vertex G along the edges of the small cubes, if we are only allowed to move in the directions of the vectors $\overrightarrow{AB}$, $\overrightarrow{AD}$ and $\overrightarrow{AE}$.

3. Let p, q, r and s be positive integers, for which

$$p^2 + 13q^2 + 25r^2 + 9s^2 = 6(pq - 2qr - 4rs)$$

holds. Prove that p is a multiple of 27.

A-I

1. Let a, b, c be arbitrary positive real numbers. Prove that the inequality

$$\left(1 + \frac{a}{b+c}\right)\left(1 + \frac{b}{c+a}\right)\left(1 + \frac{c}{a+b}\right) \geq \frac{27}{8}$$

holds. When does equality hold?

2. Determine all possible positive values of the product pqr if we are given real numbers p, q, r, for which

$$pq + q + 1 = qr + r + 1 = rp + p + 1 = 0$$

holds.

3. Let $ABCD$ be a square with sides of length a and $BPQR$ be a square with sides of length $b < a$, such that R lies on the segment BC and P lies on the extension of AB beyond B. Let X be the point of intersection of lines AR and BD. Furthermore, let Y be the point of intersection of lines BQ and PC.

(a) Prove that $AR \perp PC$ holds.
(b) Prove that $XY \parallel AB$ holds.

4. Find all pairs (x, y) of positive integers such that

$$[x, y] = [x + 3, y + 3] \quad \text{and} \quad (x + 3, y + 3) = 3(x, y)$$

hold.

Note: Here, $[x, y]$ denotes the greatest common divisor and (x, y) the least common multiple of positive integers x, y.

A-T

1. Solve the following system of equations in real numbers:

$$x + 3yz = 24,$$
$$y + 3zx = 24,$$
$$z + 3xy = 24.$$

2. Let $ABCD$ be a unit square. Arbitrary points P and Q are chosen on its sides BC and CD respectively, such that $|\angle PAQ| = 45°$ holds. Prove that the perimeter of the triangle PCQ is constant (i.e., it is independent of the specific choice of points P and Q), and determine its value.

3. Points $P_1, P_2, \ldots, P_{2006}$ are placed at random intervals on a circle, such that no two of them are in the same spot. Determine the total number of different convex polygons, whose vertices are all among the points P_i.

Duel XV: Chorzów, 2007

C-I

1. Determine whether the equation

$$x^3 - x = 7654321$$

has an integer solution.

2. Prove that only one positive real number p exists, such that the value of the fraction

$$\frac{4p}{p^2 + 4}$$

is an integer.

3. We are given an equilateral triangle ABC with sides of the length a. Points B_1, C_1 and A_1 lie on the rays AB, BC and CA beyond vertices

B, C and A, respectively, such that $|AA_1| = |BB_1| = |CC_1| = a$ holds. Determine the ratio of the areas of triangles ABC and $A_1B_1C_1$.

4. We are given a square $KLMN$ with sides of length 5 cm and an interior point P of the side LM with $|LP| = 3$ cm. Construct the points X and Y on segments KM and KP, respectively, such that $|KX| = |XY| = |YM|$ holds.

C-T

1. Determine the number of ways in which a 4×4 chessboard can be covered using eight 2×1 dominos.

2. We are given a real function $f(x) = x^2 + ax + b$ with

$$f(a) = 51, \quad f(b) = b^3 + b^2(a+1) \quad \text{and} \quad f(1) < 0.$$

Determine the values of a and b.

3. A hexagon $ABCDEF$ is inscribed in a circle with radius 3 and diameter AD. Its sides AB, CD, DE and FA all have the length 2. Determine the area and the circumference of the hexagon.

B-I

1. Determine all real values of the parameter a such that the polynomial

$$x^4 + 4x^3 + 10x^2 + 4ax + 21$$

is divisible by the polynomial $x^2 + a$.

2. We are given a square with the area P and circumcircle k. Let us consider two lines parallel to the sides of the square and passing through a common interior point of the square. These parallels intersect k in the vertices of a quadrilateral with the area Q. Prove that $P \geq Q$ must hold.

3. We are given the equation $x^2y + 3x = c$ with an integer parameter c.

(a) Determine all possible values of c such that the given equation has at least one solution (x, y) in positive integers.
(b) Prove that there exist at least two solutions (x_1, y_1) and (x_2, y_2) in integers if c is divisible by 9.

(c) Prove that there exist infinitely many values of c, for which the equation has exactly one solution (x, y) in positive integers.

4. Let $ABCDE$ be a convex pentagon with congruent interior angles at the vertices C, D and E, and in which $|BC| = |DE|$ and $|CD| = |AE|$ hold. Prove that the pentagon $ABCDE$ is cyclic.

B-T

1. Determine all triples (a, b, c) of real parameters such that the following three quadratic equations have exactly one common real root:

$$x^2 + (2a + b + c)x + a = 0,$$
$$x^2 + (2b + c + a)x + b = 0,$$
$$x^2 + (2c + a + b)x + c = 0.$$

2. We are given a square $ABCD$. Construct a rectangle $KLMN$ fulfilling the following conditions:

(a) vertices K, L, M, N lie on the sides AB, BC, CD, DA, respectively, and

(b) there exists a circle inscribed in the pentagon $KLCMN$.

3. Determine all four-digit positive integers n of the form $\overline{abba}$ such that

(a) n is a perfect square,
(b) n is a perfect cube.

A-I

1. Suppose the quadratic equation

$$x^2 - 2007x + b = 0$$

with a real parameter b has two positive integer roots. Determine the maximum value of b.

2. Determine all positive real numbers x, y and z for which the system of equations:

$$x(x + y) + z(x - y) = 65,$$
$$y(y + z) + x(y - z) = 296,$$
$$z(z + x) + y(z - x) = 104$$

is fulfilled.

3. Let M be the centre of a unit circle and A, B, C and P points on this circle such that ABM and BCM are equilateral triangles. Assume that MP is perpendicular to AM and that P lies in the opposite half-plane determined by the line AM with respect to B and C. Prove that the equality

$$|CP| \cdot (|AC| - 1) = \sqrt{2}$$

holds.

4. Let ABC be an acute-angled triangle. Let D be the foot of the altitude from C on the side AB and P and Q be the feet of the perpendiculars from D on AC and BC, respectively. Let X be the second point of intersection (beside P) of PQ with the circumcircle of the triangle ADP, and Y similarly the second point of intersection (beside Q) of PQ and the circumcircle of the triangle BDQ. The extension of the ray DX meets AC at the point U and the extension of the ray DY meets BC at the point V. Prove that the triangles ADU and BDV are similar.

A-T

1. Determine all real solutions (x, y) of the equation

$$2 \cdot 3^{2x} - 2 \cdot 3^{x+1} + 3^x \cdot 2^{y+1} + 2^{2y} - 2^{y+2} + 5 = 0.$$

2. Let ABC be a triangle with $|AC| = 2|AB|$ and let k be its circumcircle. The tangents of k in A and C intersect in P. Prove that the point of intersection of the line BP with the perpendicular bisector of BC lies on k.

3. Let $n \geq 3$ be a positive integer. Solve the following (cyclic) system of equations in non-negative real numbers:

$$\begin{aligned} 2x_1 + x_2^2 &= x_3^3 \\ 2x_2 + x_3^2 &= x_4^3 \\ &\vdots \\ 2x_{n-1} + x_n^2 &= x_1^3 \\ 2x_n + x_1^2 &= x_2^3. \end{aligned}$$

Duel XVI: Přerov, 2008

C-I

1. Determine the number of triangles with integer-length sides, such that two of the sides are of length m and n ($1 \leq m \leq n$). Solve this problem for the special values $m = 6$ and $n = 9$ and then for general values.

2. We are given two parallel lines p and q. Let us consider a set A of 13 different points such that 7 of them lie on p and the other 6 lie on q.

(a) How many line segments exist, whose endpoints are elements of the set A?

(b) How many triangles exist, whose vertices are elements of the set A?

3. Determine the smallest positive integers x and y such that the equality

$$12x = 25y^2$$

holds.

4. Prove that the three medians of a given triangle cut this triangle into six smaller triangles of the same area.

C-T

1. A line segment XY with $|XY| = 2$ is the common main diagonal of a regular hexagon and a square. Determine the area of the section common to both figures.

2. Let a, b, c be real numbers. Prove that the value of

$$V = 4(a^2 + b^2 + c^2) - [(a+b)^2 + (b+c)^2 + (c+a)^2]$$

is certainly a non-negative real number and determine all values of a, b, c for which $V = 0$ holds.

3. We are given a board composed of 16 unit squares as shown below. We wish to colour some of the cells green in such way that, no matter where we place the T-shaped tetromino on the board (with each square of the tetromino covering exactly one square on the board), at least one square of the tetromino will be on a green cell. Determine the smallest possible number of cells we must colour green and prove that this is the smallest number.

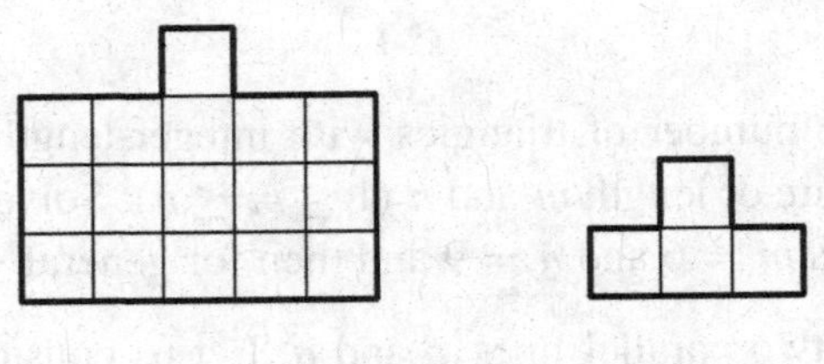

B-I

1. Show that

$$m = \frac{2008^4 + 2008^2 + 1}{2008^2 + 2008 + 1}$$

is an integer and determine its value.

2. Determine all positive integers n for which there exist positive integers x and y satisfying

$$x + y = n^2,$$
$$10x + y = n^3.$$

3. Let $ABCD$ be a trapezoid ($AB \parallel CD$) of unit area, with $|AB| = 2 \cdot |CD|$. Furthermore, let K and L be the midpoints of its sides BC and CD, respectively. Determine the area of the triangle AKL.

4. The first (lead) digit of a positive integer n is 3. If we write the same integer without the lead digit 3, we obtain an integer m.

(a) Determine all such integers n, for which $n = a \cdot m$ holds with $a = 25$.
(b) Determine two further positive integers $a \neq 25$ for which such integers n and m exist.
(c) Prove that no such integers n and m exist for $a = 32$.

B-T

1. Let $ABCD$ be a convex quadrilateral. Let K, L, M and N be the midpoints of its sides AB, BC, CD and DA, respectively.

(a) Prove that there exists a triangle with sides of lengths KL, KM and KN.

(b) Determine its area Q depending on the area P of the quadrilateral $ABCD$.

2. The arithmetic mean of 10 positive integers is 2008.

(a) What is the largest possible number among them and what is the smallest?
(b) If 304 is one of the numbers, what is largest possible number among them and what is the smallest?
(c) If we know that all 10 positive integers are different, what is the largest and smallest possible value for the 10 numbers?

3. Of the triples (4, 6, 8), (4, 8, 9) and (5, 12, 13) only one can be interpreted as giving the lengths of the three altitudes of a triangle ABC.

(a) Determine which of these triples can give the altitudes of a triangle.
(b) Determine a Euclidean construction for the triangle whose altitudes are given by the triple, and prove that the construction is complete.

A-I

1. Show that

$$n = \frac{2008^3 + 2007^3 + 3 \cdot 2008 \cdot 2007 - 1}{2009^2 + 2008^2 + 1}$$

is an integer and determine its value.

2. The orthocentre H of an acute-angled triangle ABC is reflected on the sides a, b and c yielding points A_1, B_1 and C_1, respectively. We are given $\angle C_1AB_1 = \angle CA_1B$, $\angle A_1BC_1 = \angle AB_1C$ and $\angle B_1CA_1 = \angle BC_1A$. Prove that ABC must be an equilateral triangle.

3. Let a, b, c be arbitrary positive real numbers such that $abc = 1$. Prove that the inequality

$$\frac{a}{ab+1} + \frac{b}{bc+1} + \frac{c}{ca+1} \geq \frac{3}{2}$$

holds. When does equality hold?

4. Let $ABCD$ be a tetrahedron with three mutually perpendicular edges at its vertex D. Let S denote the centre of its circumscribed sphere. Prove that the centroid T of its face ABC lies on the line DS.

A-T

1. Determine all triples (x, y, z) of positive integers such that

$$3 + x + y + z = xyz$$

holds.

2. Let D be a point on the side BC of a given triangle ABC such that

$$|AB| + |BD| = |AC| + |CD|$$

holds. The line segment AD intersects the incircle of ABC at X and Y with X closer to A. Let E be the point of tangency of the incircle of ABC with BC. Let I denote the incentre of ABC and M the midpoint of the line segment BC. Show that

(a) the line EY is perpendicular to AD and
(b) $|XD| = 2|IM|$ holds.

3. Determine all functions $f : \mathbb{Z} \to \mathbb{R}$ such that

$$f(3x - y) \cdot f(y) = 3f(x)$$

holds for all integers x and y.

Duel XVII: Graz, 2009

C-I

1. We are given a three-digit number a. The number b has the same digits as a, but in reverse order. Prove that the number $200a+9b$ is always divisible by 11.

2. Three circles with centres S_1, S_2 and S_3 are externally tangent to each other. We are given $|S_1S_2| = 8$, $|S_2S_3| = 10$ and $|S_3S_1| = 11$. Determine the radii of the three circles.

3. A *Goodword* is a string of letters, in which there is always at least one vowel between any two consonants, i.e., in which no two consonants appear next to each other. We wish to form Goodwords from the letters of the word "duel".

(a) How many different four-letter Goodwords can be formed using all four letters?

(b) How many different four-letter Goodwords can be formed with these letters if they can each be used more than once (and therefore not all letters must be used in any specific Goodwords)?

4. Let ABC be a right-angled triangle with hypotenuse length c. Determine the area of the triangle if the sum $s = a + b$ of the lengths of the sides is given. Express your result exclusively in terms of c and s.

C-T

1. (a) Determine the smallest three-digit positive integer with exactly six even divisors.

(b) Determine the largest three-digit positive integer with exactly six odd divisors. (Note that 1 is a divisor of any positive integer.)

2. We are given a line segment AB with length $|AB| = 4$. P is a point on AB and C and D are points on a semi-circle with diameter AB such that $PC \perp PD$ and $\angle CPA = \angle DPB$. Prove that the length of the segment CD is independent of the choice of P and determine the length of CD.

3. We are given a number x such that the equality $x^4 = x - 2$ holds.

(a) We know that x^{10} can be expressed as $x^{10} = ax^3 + bx^2 + cx + d$ for some integer values of a, b, c and d. Determine the values of a, b, c and d.

(b) We also know that it is possible to express x^{10} as $x^{10} = x^k + px^{k-1} + qx^{k-2}$ for some positive integer k and some integers p and q. Determine values for k, p and q.

B-I

1. Let $n \geq 2$ be an integer. Decide whether the sum of n consecutive integers can be a prime.

2. A point D lies on the side BC of an isosceles triangle ABC with base AB such that $BD = 2 \cdot DC$. Let P be the point in which the line segment AD intersects the circle with diameter BC. Prove $\angle PAC = \angle PBA$.

3. Determine all pairs of real numbers (a, b) such that the equations $x^2 + ax + b = 0$ and $x^2 + 2bx - 4a = 0$ each have two different real solutions, and the solutions of the second equation are the reciprocals of the solutions of the first.

4. (a) Determine all quadratic polynomials $P(x)$ with real coefficients, such that

$$(P(x))^2 = 2 \cdot P(x^2) - 2 \cdot P(x) + 1$$

holds.

(b) Show that no quadratic polynomial with real coefficients exists, such that

$$(P(x))^2 = 2 \cdot P(x^2) + 2 \cdot P(x) + 1$$

holds.

B-T

1. Determine the largest positive integer k such that the number

$$n^6 - n^4 - n^2 + 1$$

is divisible by 2^k for every odd integer $n > 1$.

2. We are given a right-angled triangle ABC with right angle in C. Let D be the foot of C on the hypotenuse c, and I_1 and I_2 the incentres of triangles BCD and CAD, respectively. The incircle of BCD is tangent to BC in T_1 and the incircle of CAD is tangent to CA in T_2. Prove that the lines T_1I_1 and T_2I_2 intersect on the hypotenuse AB.

3. Determine all real solutions of the equation

$$\left[\frac{x}{2}\right] + \left[\frac{x}{4}\right] + \left[\frac{x}{8}\right] = 2009 - x.$$

(Note that $[a]$ denotes the largest integer less than or equal to a.)

A-I

1. Determine all triples (x, y, z) of real numbers satisfying the following system of equations:

$$x^2 + y = z^3 + 1,$$
$$y^2 + z = x^3 + 1,$$
$$z^2 + x = y^3 + 1.$$

2. Let (a_p) be an arithmetic sequence. The conditions $a_m = 2n + m$ and $a_n = 2m + n$ are fulfilled for certain integers m and n. Express a_p in terms of p.

3. Let α and β be the interior angles of a triangle ABC at A and B, respectively, such that

$$1 + \cos^2(\alpha + \beta) = cos^2\alpha + \cos^2 \beta$$

holds. Prove that ABC is right angled.

4. A positive integer n is called *notable*, if it can be expressed in the form $n = 7x^2 + 287y^2$, with non-negative integer values for x and y. Prove that both $2009 \cdot n$ and $2010 \cdot n$ are notable for all notable values of n.

A-T

1. Let $a_1, a_2, \ldots, a_5$ be positive real numbers. Prove

$$\frac{a_1 + a_2}{a_3 + a_4 + a_5} + \frac{a_2 + a_3}{a_4 + a_5 + a_1} + \frac{a_3 + a_4}{a_5 + a_1 + a_2} + \frac{a_4 + a_5}{a_1 + a_2 + a_3} + \frac{a_5 + a_1}{a_2 + a_3 + a_4} \geq \frac{10}{3}.$$

When does equality hold?

2. Let $\rho = 2$ be the inradius and ρ_a, ρ_b and ρ_c the radii of the excircles of a triangle ABC. Determine the area of ABC, if ρ_a, ρ_b and ρ_c are integers.

3. Some of the cells of an $n \times n$ chessboard are coloured black. Determine the smallest possible number of cells that must be black, such that each 2×2 square on the chessboard contains at least two black cells.

Duel XVIII: Chorzów, 2010

C-I

1. Prove that the sum of eight consecutive odd positive integers is divisible by the number 8.

2. We are given a rectangle $ABCD$. Arbitrary triangles ABX and CDY are erected on its sides AB and CD. We define the midpoints P of AX, Q of BX, R of CY and S of DY. Prove that the line segments PR and QS have a common midpoint.

3. Let us consider a trapezoid with sides of lengths 3, 3, 3, k, with positive integer k. Determine the maximum area of such a trapezoid.

4. Prove that each positive integer $n \geq 6$ can be written as a sum of two positive integers, one of which is prime and the second of which is a composite number.

C-T

1. Determine the number of pairs (x, y) of decimal digits such that the positive integer in the form $\overline{xyx}$ is divisible by 3 and the positive integer in the form $\overline{yxy}$ is divisible by 4.

2. We are given a cube $ABCDEFGH$ with edges of the length 5 cm. Determine the lengths of all altitudes of the triangle EDC.

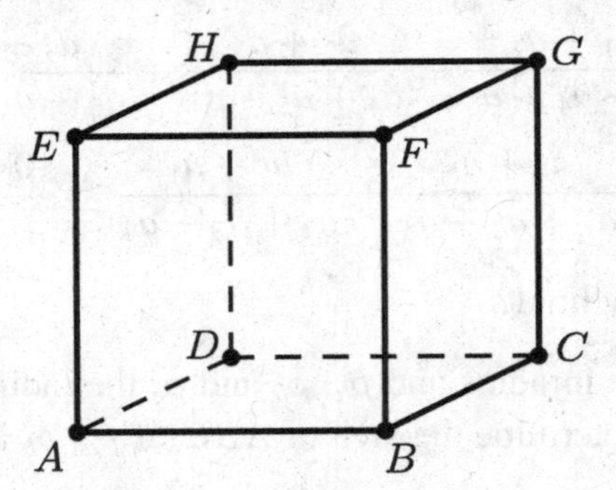

3. The sum of the digits of a three digit prime p_1 is a two-digit prime p_2. The sum of the digits of p_2 is a one-digit prime $p_3 > 2$. Determine all triples (p_1, p_2, p_3) of such primes.

B-I

1. Prove that the number $2010^{2011} - 2010$ is divisible by $2010^2 + 2011$.

2. Determine all pairs (x, y) of positive integers such that the equation

$$4^x = y^2 + 7$$

is fulfilled.

3. We are given two line segments AB and CD in the plane. Determine the set of all points V in this plane such that the triangles ABV and CDV have the same area.

4. Determine all polynomials $P(x)$ with real coefficients and all real numbers q such that the equation

$$2xP(x-2) - P(x^2) = 3x^2 - 22x + q$$

holds for all real numbers x.

B-T

1. Let us consider the triangle ABC with altitudes $h_a = 24$ and $h_b = 32$. Prove that its third altitude h_c fulfils the inequalities

$$13 < h_c < 96.$$

2. A square piece of paper $ABCD$ is folded such that the corner A comes to lie on the midpoint M of the side BC. The resulting crease intersects AB in X and CD in Y. Show that $|AX| = 5|DY|$.

3. How many positive integers of the form $\overline{abcabcabc}$ exist that are divisible by 29?

A-I

1. Determine all triples of mutually distinct real numbers a, b, c such that the cubic equation

$$x^3 + abx^2 + bcx + ca = 0$$

with unknown x has three real roots a, b and c.

2. We are given bases $|AB| = 23$ and $|CD| = 5$ of a trapezoid $ABCD$ with diagonals $|AC| = 25$ and $|BD| = 17$. Determine the lengths of its sides BC and AD.

3. We are given a circle c_1 and points X and Y on c_1. Let XY be a diameter of a second circle c_2. We choose a point P on the greater arc XY on c_1 and a point Q on c_2 such that the quadrilateral $PXQY$ is convex and $PX \parallel QY$ holds. Prove that the size of the angle PYQ is independent of the choice of P (if an appropriate Q exists).

4. Solve the following system of equations in real numbers

$$\sqrt{\sqrt{x} + 2} = y - 2,$$

$$\sqrt{\sqrt{y} + 2} = x - 2.$$

A-T

1. We are given two real numbers x and y ($x \neq y$) such that $x^4 + 5x^3 = y$ and $y^3 + 5x^2 = 1$ hold. Prove that the equality $x^3 + x^2y + xy^2 = -1$ holds.

2. Let us consider a unit square $ABCD$. On its sides BC and CD, determine points E and F respectively, such that $|BE| = |DF|$ holds, and the triangles ABE and AEF have the same perimeter.

3. Determine all pairs (x, y) of integers such that the equation

$$4^y + 1899 + x^3$$

is fulfilled.

Duel XIX: Přerov, 2011

C-I

1. The number n has the following properties:

(a) the product of all its digits is odd,
(b) the sum of the squares of its digits is even.

Prove that the number of digits in n cannot be equal to 2011.

2. Let ABC be a right triangle with hypotenuse AB. Determine measures of its angles A and B if the angle bisector at B divides the opposite side AC at a point D such that $|AD|:|CD| = 2:1$ holds.

3. Determine the number of ten-digit numbers divisible by 4, which are written using only the digits 1 and 2.

4. Let p, q be two parallel lines and A a point lying outside the strip bounded by p and q. Construct a square $ABCD$ such that its vertices B and D lie on p and q, respectively.

C-T

1. Determine all pairs (x, y) of positive integers satisfying the equation

$$(x + y)^2 = 109 + xy.$$

2. We are given an isosceles triangle ABC with the base $|AB| = \sqrt{128}$. The foot of its altitude from A divides the side BC into two parts in the ratio 1:3. Determine the perimeter and the area of the triangle.

3. Determine all positive integers n such that the number $n^3 - n$ is divisible by 48.

B-I

1. Let A be a six-digit positive integer which is formed using only the two digits x and y. Furthermore, let B be the six-digit integer resulting from A if all digits x are replaced by y and simultaneously all digits y are replaced by x. Prove that the sum $A + B$ is divisible by 91.

2. We are given an isosceles right triangle EBC with the right angle at C and $|BC| = 2$. Determine all possible values of the area of a trapezoid $ABCD$ ($AB \parallel CD$) in which E is the midpoint of AD.

3. Prove that there exist infinitely many solutions of the equation

$$2^x + 2^{x+3} = y^2$$

in the domain of positive integers.

4. We are given a common external tangent t to circles $c_1(O_1, r_1)$ and $c_2(O_2, r_2)$, which have no common point and lie in the same half-plane defined by t. Let d be the distance between the tangent points of circles c_1 and c_2 with t. Determine the smallest possible length of a broken line AXB (i.e., the union of line segments AX and XB) such that A lies on c_1, B lies on c_2, and X lies on t.

B-T

1. Solve the following equation in the domain of positive integers:

$$\frac{2}{x^2} + \frac{3}{xy} + \frac{4}{y^2} = 1.$$

2. Let E be the midpoint of the side CD of a convex plane quadrilateral $ABCD$. We are given that the area of AEB is half of the area of $ABCD$. Prove that $ABCD$ must be a trapezoid.

3. Determine all real solutions of the following system of equations:

$$2a - 2b = 29 + 4ab,$$
$$2c - 2b = 11 + 4bc,$$
$$2c + 2a = 9 - 4ac.$$

A-I

1. Solve the following system of equations in the domain of real numbers:

$$\begin{aligned} x^4 + 1 &= 2yz, \\ y^4 + 1 &= 2zx, \\ z^4 + 1 &= 2xy. \end{aligned}$$

2. We are given a trapezoid $ABCD$ with $AB \parallel CD$ and $|AB| = 2|CD|$. Let M be the common point of the diagonals AC and BD and E the midpoint of AD. Lines EM and CD intersect in P. Prove that $|CP| = |CD|$ holds.

3. Let a, b, p, q and $p\sqrt{a} + q\sqrt{b}$ be positive rational numbers. Prove that the numbers $\sqrt{a}$ and $\sqrt{b}$ are also rational.

4. We are given an acute-angled triangle ABC. We consider the triangle KLM with vertices in the feet of the altitudes of the given triangle. Prove that the orthocentre of ABC is also the incentre of KLM.

A-T

1. Determine all polynomials $f(x) = x^2 + px + q$ with integer coefficients p, q such that $f(x)$ is a perfect square for infinitely many integers x.

2. Let c be a circle with centre O and radius r and l a line containing O. Furthermore, let P and Q be points on c symmetric with respect to l. Also, let X be a point on c such that $OX \perp l$ and A, B be the points of intersection of XP with l and XQ with l, respectively. Prove that $|OA| \cdot |OB| = r^2$ holds.

3. Peter throws two dice simultaneously and then writes the total number of dots showing on a blackboard. Find the smallest number k with the following property: After k throws, Peter can always choose some of the written numbers, such that their product leaves the remainder 1 after division by 13.

Duel XX: Bílovec, 2012

C-I

1. Determine all positive integers x, such that

$$\frac{x}{2} + \frac{2}{x}$$

is an integer.

2. We are given a trapezoid $ABCD$ with $AB \parallel CD$, such that a point E exists on the side BC with $|CE| = |CD|$ and $|BE| = |AB|$. Prove that AED is a right triangle.

3. Two positive integers are called friends if each is composed of the same number of digits, the digits in one are in increasing order and the digits in the other are in decreasing order, and the two numbers have no digits in common (like for example the numbers 147 and 952).

(a) Determine the number of all two-digit numbers that have a friend.
(b) Determine the largest number that has a friend.

4. Let ABC be a right-angled triangle with the hypotenuse AB such that $|AC|:|BC| = 2:3$. Let D be the foot of its altitude from C. Determine the ratio $|AD|:|BD|$.

C-T

1. Determine the number of seven-digit numbers divisible by 4, with the property that the sum of their digits is equal to 4.

2. We are given a right triangle ABC with right angle at C. A point D lies on AB such that $|BD| = |BC|$. A point E lies on the line perpendicular to AB and passing through A such that $|AE| = |AC|$. The points E and C are in the same half-plane defined by AB. Show that the points C, D and E lie on a common line.

3. We are given eight coins, no two of which have the same weight, and a scale with which we can determine which group of coins placed on either end is heavier and which is lighter. We wish to determine which of the eight coins is the heaviest and which is the lightest. Prove that this can be done with at most 10 weighings.

B-I

1. Determine all pairs (p, x) fulfilling the equation in positive integers

$$x^2 = p^3 + 1,$$

where p is a prime and x is an integer.

2. We are given a parallelogram $ABCD$. A line l passing through B meets the side CD at the point E and the ray AD at the point F. Determine the ratio of the areas of the triangles ABF and BEC in terms of the ratio $|CE|:|ED|$.

3. Let k and n be arbitrary real numbers with $1 \leq k \leq n$. Prove that the inequality

$$k(n - k + 1) \geq n$$

holds. When does equality hold?

4. Prove that 2012 cannot be written as the sum of two perfect cubes. It is possible to write 2012 as the difference of two perfect cubes? If not, prove that it is impossible.

B-T

1. Determine all real polynomials $P(x)$ such that

$$P(P(x)) = x^4 + ax^2 + 2a$$

holds for some real number a.

2. We are given an isosceles right triangle ABC. Let K be the midpoint of its hypotenuse AB. Find the set of vertices L of all isosceles right triangles KLM with hypotenuse KL such that M lies on the side AC.

3. Determine all triples (x, y, z) of positive integers for which each of the three numbers x, y, z is a divisor of the sum $x + y + z$.

A-I

1. Solve the following system of equations in the domain of integers:

$$x + \frac{2}{y} = z,$$
$$y + \frac{4}{z} = x,$$
$$z - \frac{6}{x} = y.$$

2. We are given a cyclic quadrilateral $ABCD$ with $\angle BDC = \angle CAD$ and $|AB| = |AD|$. Prove that there exists a circle, which is tangent to all four sides of the quadrilateral $ABCD$.

3. Determine all cubic polynomials $P(x)$ with real coefficients such that the equation $P(x) = 0$ has three real roots (not necessarily different) fulfilling the following conditions:

(a) The number 1 is a root of the considered equation.
(b) For each root t of the equation $P(x) = 0$ the condition $P(2t) = t$ holds.

4. Determine the minimum value of the expression

$$V = \frac{\sin\alpha}{\sin\beta\sin\gamma} + \frac{\sin\beta}{\sin\gamma\sin\alpha} + \frac{\sin\gamma}{\sin\alpha\sin\beta},$$

where α, β, γ are interior angles of a triangle.

A-T

1. Solve the following equation in positive integers:

$$xyz = 2x + 3y + 5z.$$

2. Let us consider an acute-angled triangle ABC. Let D, E, F be the feet of the altitudes from the vertices A, B, C, respectively. Furthermore, let K, L, M denote the points of intersection of the lines AD, BE, CF with the circumcircle of the triangle ABC different from the vertices A, B, C, respectively. Prove that the following inequality holds:

$$\min\left\{\frac{|KD|}{|AD|}, \frac{|LE|}{|BE|}, \frac{|MF|}{|CF|}\right\} \leq \frac{1}{3}.$$

3. Peter has a geometry kit containing six sticks of the same length but in six different colours. He can use these to construct a regular tetrahedron. How many distinct tetrahedra can he construct?

Duel XXI: Graz, 2013

C-I

1. Let $ABCD$ be a parallelogram. The circle c with diameter AB passes through the midpoint of the side CD and through the point D. Determine the measure of the angle $\angle ABC$.

2. Let n be a positive integer. Prove that the number 10^n can always be written as the sum of the squares of two different positive integers.

3. Joe is travelling by train at a constant speed v. Every time the train passes over a weld seam in the tracks, he hears a click. The weld seams are always exactly 15 m apart. If Joe counts the number of clicks, how many seconds must he count until the number of clicks is equal to the speed of the train in km/h?

4. Determine all three-digit numbers that are exactly 34 times as large as the sum of their digits.

C-T

1. In a triangle ABC with $|AB| = 21$ and $|AC| = 20$, points D and E are chosen on segments AB and AC, respectively, with $|AD| = 10$ and $|AE| = 8$. We find that AC is perpendicular to DE. Calculate the length of BC.

2. We consider positive integers that are written in decimal notation using only one digit (possibly more than once), and call such numbers *uni-digit numbers*.

(a) Determine a uni-digit number written with only the digit 7 that is divisible by 3.

(b) Determine a uni-digit number written with only the digit 3 that is divisible by 7.

(c) Determine a uni-digit number written with only the digit 5 that is divisible by 7.

(d) Prove that there cannot exist a uni-digit number written with only the digit 7 that is divisible by 5.

3. We are given a circle c_1 with midpoint M_1 and radius r_1 and a second circle c_2 with midpoint M_2 and radius r_2. A line t_1 through M_1 is tangent to c_2 in P_2 and a line t_2 through M_2 is tangent to c_1 in P_1. The line t_1 intersects c_1 in Q_1 and the line t_2 intersects c_2 in Q_2 in such a way that the points P_1, P_2, Q_1 and Q_2 all lie on the same side of M_1M_2. Prove that the lines M_1M_2 and Q_1Q_2 are parallel.

B-I

1. (a) Determine all positive integers n such that the number

$$n^4 + 2n^3 + 2n^2 + 2n + 1$$

is a prime.

(b) Determine all positive integers n, such that the number

$$n^4 + 2n^3 + 3n^2 + 2n + 1$$

is a prime.

2. Two circles c_1 and c_2 with radii r_1 and r_2 respectively ($r_1 > r_2$) are externally tangent in point C. A common external tangent t of the two circles is tangent to c_1 in A and to c_2 in B. The common tangent of the two circles in C intersects t in the midpoint of AB. Determine the lengths of the sides of triangle ABC in terms of r_1 and r_2.

3. Let s_n denote the sum of the digits of a positive integer n. Determine whether there are infinitely many integers that cannot be represented in the form $n \cdot s_n$.

4. We call a number that is written using only the digit 1 in decimal notation a *onesy* number, and a number using only the digit 7 in decimal notation a *sevensy* number. Determine a onesy number divisible by 7 and prove that for any sevensy number k, there always exists a onesy number m, such that m is a multiple of k.

B-T

1. Two lines p and q intersect in a point V. The line p is tangent to a circle k in the point A. The line q intersects k in the points B and C. The angle bisector of $\angle AVB$ intersects the segments AB and AC in the points K and L, respectively. Prove that the triangle KLA is isosceles.

2. Determine all integer solutions (x, y) of the equation

$$\frac{2}{x} + \frac{3}{y} = 1.$$

3. We are given a function $f : \mathbb{R} \to \mathbb{R}$ such that $f(m+n) = f(m)f(n)$ holds for all real values of m and n. Furthermore, we know that $f(8) = 6561$.

(a) Prove that there exists exactly one real k such that $f(k) = \frac{1}{3}$ and determine the value of k.

(b) Prove that no real number l exists such that $f(l) = -\frac{1}{3}$ holds.

A-I

1. Let a be an arbitrary real number. Prove that real numbers b and c certainly exist such that

$$\sqrt{a^2 + b^2 + c^2} = a + b + c.$$

2. Let $\mathbb{R}^+ = (0; +\infty)$. Determine all functions $f : \mathbb{R}^+ \to \mathbb{R}$ such that

$$xf(x) = xf\left(\frac{x}{y}\right) + yf(y)$$

holds for all positive real values x and y.

3. Let O be the circumcentre of an acute-angled triangle ABC. Let D be the foot of the altitude from A to the side BC. Prove that the angle bisector $\angle CAB$ is also the bisector of $\angle DAO$.

4. Let α, β, γ be the interior angles of a triangle with $\gamma > 90°$. Prove that the inequality

$$\tan\alpha \tan\beta < 1$$

holds.

A-T

1. We are given the following system of equations:

$$x + y + z = a,$$
$$x^2 + y^2 + z^2 = b^2$$

with real parameters a and b. Prove that the system of equations has a solution in real numbers if and only if the inequality

$$|a| \le |b|\sqrt{3}$$

holds.

2. We are given positive real numbers x, y, z, u with $xyzu = 1$. Prove

$$\frac{x^3}{y^3} + \frac{y^3}{z^3} + \frac{z^3}{u^3} + \frac{u^3}{x^3} \ge x^2 + y^2 + z^2 + u^2.$$

When does equality hold?

3. We call positive integers that are written in decimal notation using only the digits 1 and 2 *Graz numbers*. Note that 2 is a one-digit Graz number divisible by 2^1, 12 is a Graz number divisible by 2^2 and 112 is a three-digit Graz number divisible by 2^3.

(a) Determine the smallest four-digit Graz number divisible by 2^4.
(b) Determine an n-digit Graz number divisible by 2^n for $n > 4$.
(c) Prove that there must always exist an n-digit Graz number divisible by 2^n for any positive integer n.

Duel XXII: Přerov, 2014

C-I

1. Determine the smallest positive integer that can be expressed as the sum of two perfect squares in two ways. (Note that $10 = 3^2 + 1^2$ is one way of expressing the number 10 as the sum of two perfect squares.)

2. We are given a regular hexagon $ABCDEF$ with the area P. Lines CD and EF intersect at the point G. Determine the areas of the triangles ABG and BCG in terms of P.

3. We are given positive integers m and n such that $m^5 + n^5 = 7901$ holds. Determine the value of the expression $m^3 + n^3$.

4. Let $A = 2x83$, $B = 19y6$, $C = 29x6$ and $D = 1y54$ be four-digit numbers. Determine all pairs of decimal digits (x, y) such that both numbers $A + B$ and $C - D$ are divisible by 3.

C-T

1. We are given a 4×4 array consisting of 16 unit squares. Determine the number of ways in which the array can be covered with five congruent straight triominoes (3×1 rectangles) such that exactly one unit square in the array remains empty.

2. Two circles $k_1(M_1; r_1)$ and $k_2(M_2; r_2)$ intersect in points S and T. The line M_1M_2 intersects k_1 at points A and B and k_2 in C and D such that B lies in the interior of k_2 and C lies in the interior of k_1. Prove that the lines SC and SB trisect the angle ASD if and only if $\angle M_1SM_2 = 90°$.

3. The sum of the squares of four (not necessarily different) positive integers a, b, c, d is equal to 100, i.e.,

$$a^2 + b^2 + c^2 + d^2 = 100, \quad \text{with } a \geq b \geq c \geq d > 0.$$

(a) Determine the largest possible positive value of $a - d$. Explain why a larger value cannot be possible.

(b) Determine the smallest possible positive value of $a - d$. Explain why a smaller value cannot be possible.

B-I

1. Determine all integers n such that the fraction

$$\frac{8n - 1}{11n - 2}$$

is reducible.

2. Let x, y be real numbers with $xy = 4$. Determine the smallest possible value of the expression

$$U = |x^3 + x^2y + xy^2 + y^3|.$$

Determine all pairs of real numbers (x, y) for which the expression U achieves this smallest value.

3. Prove that the sum of the lengths of two segments connecting the midpoints of opposite sides of an arbitrary quadrilateral is less than the sum of the lengths of its diagonals.

4. One number from the set $\{-1, 1\}$ is written in each vertex of a square. In one step, the number in each vertex is replaced by the product of this number and the two numbers in adjacent vertices. Prove that it is possible to reach a situation in which the number 1 is written in each vertex after a certain number of steps if and only if the number 1 is already written in all four vertices from the start.

B-T

1. Determine all integer solutions of the equation

$$x^6 = y^3 + 4069.$$

2. Let ABC be an acute-angled triangle. Let E be the perpendicular foot of A on BC and F the perpendicular foot of B on AC. Furthermore, let M be the midpoint of BE and N the midpoint of AF. Prove that the line through M perpendicular to AC and the line through N perpendicular to BC intersect in the midpoint of EF.

3. Find all real solutions of the following system of equations:

$$2a^2 = b^2 - \frac{1}{c^2} + 2,$$

$$2b^2 = c^2 - \frac{1}{a^2} + 2,$$

$$2c^2 = a^2 - \frac{1}{b^2} + 2.$$

A-I

1. Let $a \neq 0, b, c$ be real numbers with $|a + c| < |a - c|$, Prove that the quadratic equation $ax^2 + bx + c = 0$ has two real roots, one of which is positive and one of which is negative.

2. We are given a tetrahedron $ABCD$ with pairwise perpendicular edges at vertex D. Let K, L, M be the midpoints of edges AB, BC, CA, respectively. Prove that the sum of the measures of the angles at vertex D in three adjacent faces of the tetrahedron $KLMD$ is equal to $180°$.

3. Let $f : \mathbb{R} \to \mathbb{R}$ be a real function satisfying

$$f(f(x) - y) = x - f(y) \quad \text{for all } x, y \in \mathbb{R}.$$

Prove that f is an odd function, i.e., $f(-x) = -f(x)$ holds for all $x \in \mathbb{R}$.

4. Determine all positive integers n for which there exist mutually distinct positive integers $a_1, a_2, \ldots, a_n$ such that

$$\frac{1}{a_1} + \frac{1}{a_2} + \cdots + \frac{1}{a_n} = 1.$$

A-T

1. Prove that the inequality

$$(a+9)\left(a^2 + \frac{1}{a} + \frac{b^2}{8}\right) \geq (a+b+1)^2$$

holds for all positive real values a and b. When does equality hold?

2. Red, green and blue fireflies live on the magic meadow.

- If two blue fireflies meet, they change into one red firefly.
- If three red fireflies meet, they change into one blue firefly.
- If three green fireflies meet, they change into one red firefly.
- If one red and one green firefly meet, they change into one blue firefly.
- If one red and one blue firefly meet, they change into two green fireflies.

Initially, there are 2014 red fireflies on the meadow. At some point, there are exactly five fireflies on the meadow. Prove that these five fireflies do not all have the same colour. Is it possible for exactly one firefly to remain on the meadow?

3. We are given seven distinct positive integers. Prove that four of these can be chosen, such that their sum is divisible by four.

Duel XXIII: Bielsko Biała, 2015

C-I

1. Determine all pairs (m, n) of integers satisfying the following equation:

$$m + \frac{1}{n} = n + \frac{1}{m}.$$

2. Let ABC be an acute-angled triangle with integer angles α, β and γ. We are given that the external angle ε at vertex A is an integer multiple of α with $\varepsilon = k\alpha$. Determine all possible values of k.

3. Determine all triples (a, b, c) of positive integers such that each of the following expressions has an integer value:

$$\frac{a+b}{b+c}, \quad \frac{b+c}{c+a}, \quad \frac{c+a}{a+b}.$$

4. We are given a rectangle $ABCD$ with $|AB| = 4$ and $|\angle ABD| = 30°$. A point E lies on the circumcircle of $ABCD$ with $CE \parallel BD$. Determine the length of segment AE.

C-T

1. Determine the number of pairs of integers (x, y) for which the following inequality is fulfilled:

$$2|x| + 3|y| < 23.$$

2. We are given a right triangle ABC with legs $|AC| = 1$ and $|BC| = \sqrt{3}$. Let us consider the circles with diameters AC and BC. Calculate the area of the section common to both circles.

3. A *wavy number* is a number in which the digits alternately get larger and smaller (or smaller and larger) when read from left to right. (For instance, 3 629 263 and 84 759 are wavy numbers but 45 632 is not.)

(a) Two five-digit wavy numbers m and n are composed of all digits from 0 to 9. (Note that the first digit of a number cannot be 0.) Determine the smallest possible value of $m + n$.

(b) Determine the largest possible wavy number in which no digit occurs twice.

(c) Determine a five-digit wavy number that can be expressed in the form $ab + c$, where a, b and c are all three-digit wavy numbers.

B-I

1. How many triples of positive integers (a, b, c) of positive integers with

$$abc = 45\,000$$

exist?

2. Let ABC be a triangle with right angle at vertex C. Let ACP and BCQ be right isosceles triangles external to ABC with right angles at P and Q, respectively. Furthermore, let F be the perpendicular foot of C on AB, and D and E be the points of intersection of the line AC with PF and the line BC with QF, respectively. Prove that $|DC| = |EC|$.

3. Let p, q, r, s be non-negative real numbers with $p \leq q \leq r \leq s$. Prove that the inequality

$$\frac{p+q+r+s}{4} \geq \frac{p+q+r}{3}$$

holds. When does equality hold?

4. Let $ABCD$ be a circumscribed quadrilateral with right angles at B and D. Prove that $ABCD$ is a deltoid.

B-T

1. Determine all 5-tuples (a, b, c, d, e) of positive integers, such that each of the following expressions has an integer value:

$$\frac{a+b}{c+d}, \quad \frac{b+c}{d+e}, \quad \frac{c+d}{e+a}, \quad \frac{d+e}{a+b}, \quad \frac{e+a}{b+c}.$$

2. (a) Jacek has four sticks of integer length that he puts on the table to form a convex quadrilateral. No matter which three of the four sticks he chooses, there is never any way he can form a triangle. What is the smallest possible circumference of the quadrilateral that Jacek can make?

(b) Józef has six sticks of integer length. He can put them on the table and form a convex hexagon, but just like Jacek, there is never any way he can form a triangle with three of his sticks. What is smallest possible length of the longest of Józef's sticks?

3. Determine the number of all six-digit palindromes that are divisible by seven.

A-I

1. Determine all pairs of real numbers (a, b) such that the roots of the cubic equation

$$x^3 + ax^2 + bx + ab = 0$$

are the numbers $-a$, $-b$ and $-ab$.

2. We are given an arbitrary positive integer n. Prove that a perfect square exists, such that the sum of its digits equals n^2.

3. Let u and v be the distances of an arbitrary point on the side AB of an acute-angled triangle ABC to its sides AC and BC. Furthermore, let h_a and h_b be the lengths of the altitudes from its vertices A and B, respectively. Prove that the following inequalities hold:

$$\min\{h_a, h_b\} \leq u + v \leq \max\{h_a, h_b\}.$$

4. The positive integers k, l, m, n fulfil the equation

$$k^2l^2 - m^2n^2 = 2015 + l^2m^2 - k^2n^2.$$

Find all possible values of $k + l + m + n$.

A-T

1. Determine all pairs (x, y) of integers fulfilling the equation

$$x^2 - 3x - 4xy - 2y + 4y^2 + 4 = 0.$$

2. We are given a triangle ABC. Prove that for any triple u, v, w of positive real numbers, there exists a point P inside the triangle ABC such that

$$S_{ABP}:S_{BCP}:S_{CAP} = u:v:w.$$

(Note that S_{XYZ} denotes the area of a triangle XYZ.)

3. As shown in the figure, a circle is surrounded by six touching circles of the same size.

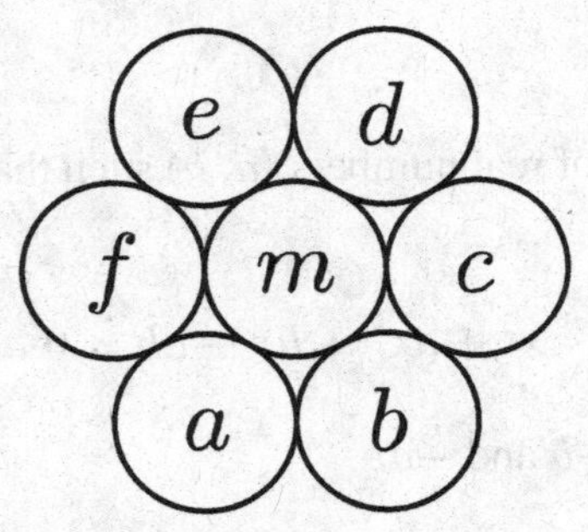

A number a, b, c, d, e, f or m is written in the interior of each circle. It is known that each of these numbers is equal to the product of all numbers in the interiors of the touching circles. Determine all possible values of m and prove that no other value is possible.

Duel XXIV: Ostrava, 2016

C-I

1. The inequality $a^2 + 2b^2 \geq pab$ is known to hold for all real values of a and b.

(a) Determine any value of p such that this is true and prove why this is true.

(b) Does this inequality hold for all real numbers a, b, provided $p = 2$ and $p = 3$? Why?

2. Each vertex of a regular hexagon $ABCDEF$ is coloured in one of three colours (red, white or blue) such that each colour is used exactly twice. Determine in how many ways we can do this, if any two adjacent vertices of the hexagon are always coloured in distinct colours.

3. We are given an equilateral triangle ABC in the plane. Let D be an interior point of the side AC. The point E lies on the ray BC beyond the point C such that $|AD| = |CE|$ holds. Prove that $|BD| = |DE|$ must hold.

4. We are given two positive real numbers x and y. Their arithmetic mean $A = (x + y)/2$ and their geometric mean $G = \sqrt{xy}$ are in the ratio $A{:}G = 5{:}4$. Determine the ratio $x{:}y$.

C-T

1. Let a, b, c be arbitrary non-zero real numbers. Let us denote

$$A = \frac{a^2+b^2}{c^2}, \quad B = \frac{b^2+c^2}{a^2}, \quad C = \frac{c^2+a^2}{b^2},$$

and furthermore $P = A \cdot B \cdot C$ and $S = A + B + C$. Determine all possible values of the difference $P - S$.

2. We are given two circles c_1 and c_2 with midpoints M_1 and M_2, respectively. The radius of c_1 is r and the radius of c_2 is $2r$. Furthermore, we are given that $|M_1M_2| = r\sqrt{2}$. Let P be a point on c_2 with the property that the tangents from P to c_1 are mutually perpendicular. Determine the area of the triangle M_1M_2P in terms of r.

3. We are given the interesting number 5040. Like some other numbers, this number can be expressed as the product of several consecutive integers. (For example, the number 1320 can be expressed in the form $1320 = 10 \cdot 11 \cdot 12$ as the product of three consecutive integers.)

(a) Express 5040 as the product of 4 consecutive integers.
(b) Express 5040 as the product of 6 consecutive integers.
(c) Prove that 5040 cannot be expressed as the product of two consecutive integers.

B-I

1. Let a, b, c be arbitrary real numbers. Prove that the inequality

$$a^2 + 5b^2 + 4c^2 \geq 4(ab + bc)$$

holds. When does equality hold?

2. Determine in how many ways one can assign numbers of the set $\{1, 2, \ldots, 8\}$ to the vertices of a cube $ABCDEFGH$ such that the sum of any two numbers at vertices with a common edge is always an odd number.

3. A circle meets each side of a rectangle at two points. Intersection points lying on opposite sides are vertices of two trapezoids. Prove that points of intersection of these two trapezoids lying inside the rectangle are vertices of a cyclic quadrilateral.

4. Consider the numbers from the set $\{1, 2, 3, \ldots, 2016\}$ How many of these have the property that its square leaves a remainder of 1 after division by 2016?

B-T

1. Determine all four-digit palindromic numbers n (i.e., numbers that read the same from front to back and from back to front) such that $17n$ is a perfect square.

2. We are given a right triangle ABC. Its legs BC and AC are simultaneously the hypotenuses of two right-angled isosceles triangles BCP and ACQ erected outside ABC. Let D be the vertex of the right-angled isosceles triangle ABD with hypotenuse AB erected inside ABC. Prove that the point D lies on the line PQ.

3. A set A consists exclusively of positive integers not divisible by 7. How many elements must the set A at least contain in order to be sure that there is a non-empty subset of A, such that the sum of the squares of the elements of this subset is divisible by 7?

A-I

1. Find the largest positive integer n with the following property: The product

$$(k+1)\cdot(k+2)\cdot(k+3)\cdot\ldots\cdot(k+2016)$$

is divisible by 2016^n for every positive integer k.

2. Solve the following system of equations in real numbers with non-negative real parameters p, q, r satisfying $p+q+r=\frac{3}{2}$:

$$x = p^2 + y^2,$$
$$y = q^2 + z^2,$$
$$z = r^2 + x^2.$$

3. We are given a square $ABCD$. Determine the locus of vertices P of all right isosceles triangles APQ with the right angle at P, such that the vertex Q lies on the side CD of the given square.

4. Exactly one flea is sitting on every square of a 10×10 array. At a signal all fleas jump diagonally over one square onto another square of the array. There are then several squares containing fleas (perhaps more than one) and others that are empty. Determine the smallest possible number of empty squares.

A-T

1. Prove that for every positive integer n, there exists an integer m divisible by 5^n which consists exclusively of odd digits.

2. A cube F with edge length 9 is divided into 9^3 small cubes with edges of the length 1 by planes parallel to its faces. One small cube is removed from the centre of each of the six faces of the cube F. Let us denote the resulting solid by G. Is it possible to build the solid G only from rectangular cuboids with the dimensions $1 \times 1 \times 3$?

3. Determine all polynomials $P(x)$ with real coefficients, such that for each real number x the equation

$$P(x^2) - 3x^3 + 15x^2 - 24x + 12 = P(x)\,P(2x)$$

holds.

Printed in the United States
By Bookmasters